EQUITY, DIVERSITY AND INT.

Perspectives on Rural Policy and Planning

Series Editors:
Andrew Gilg
University of Exeter, UK
Keith Hoggart
King's College London, UK
Henry Buller
Cheltenham College of Higher Education, UK
Owen Furuseth
University of North Carolina, USA
Mark Lapping
University of South Maine, USA

Other titles in the series

Women in the European Countryside
Edited by Henry Buller and Keith Hoggart
ISBN 0 7546 3946 0

Mapping the Rural Problem in the Baltic Countryside
Transition Processes in the Rural Areas of Estonia, Latvia and Lithuania
Edited by Ilkka Alanen
ISBN 0 7546 3434 5

Geographies of Rural Cultures and Societies
Edited by Lewis Holloway and Moya Kneafsey
ISBN 0 7546 3571 6

Big Places, Big Plans
Edited by Mark B. Lapping and Owen J. Furuseth
ISBN 0 7546 3586 4

Young People in Rural Areas of Europe
Edited by Birgit Jentsch and Mark Shucksmith
ISBN 0 7546 3478 7

Power and Gender in European Rural Development
Edited by Henri Goverde, Henk de Haan and Mireia Baylina
ISBN 0 7546 4020 5

Equity, Diversity and Interdependence
Reconnecting Governance and People through Authentic Dialogue

MICHAEL MURRAY and BRENDAN MURTAGH
Queen's University, Belfast, UK

Routledge
Taylor & Francis Group

LONDON AND NEW YORK

First published 2004 by Ashgate Publishing

Reissued 2018 by Routledge
2 Park Square, Milton Park, Abingdon, Oxon OX14 4RN
711 Third Avenue, New York, NY 10017, USA

Routledge is an imprint of the Taylor & Francis Group, an informa business

First issued in paperback 2018

Publisher's Note
The publisher has gone to great lengths to ensure the quality of this reprint but points out that some imperfections in the original copies may be apparent.

Disclaimer
The publisher has made every effort to trace copyright holders and welcomes correspondence from those they have been unable to contact.

ISBN 13: 978-0-815-38884-5 (hbk)
ISBN 13: 978-1-138-61958-6 (pbk)
ISBN 13: 978-1-351-15892-3 (ebk)

Contents

List of Figures   *vii*
List of Tables   *viii*
The Authors   *ix*
Preface   *xi*
Acknowledgements   *xvii*

PART ONE: SETTING THE CONTEXT

1   Discourse, Planning and Place   3

2   Introducing Equity, Diversity and Interdependence   15

**PART TWO: BASELINING EQUITY, DIVERSITY AND
INTERDEPENDENCE IN RURAL NORTHERN IRELAND**

3   The Governance Arena of Rural Northern Ireland   29

4   Perceptions of Diversity and Inclusion Among Northern
Ireland-wide Service Organisations   43

5   Perceptions of Diversity and Inclusion Among Northern Ireland
Area-based Service Organisations   81

6   Responding to the Challenges of Diversity and Inclusion –
Northern Ireland Rural Community Network   105

7   A Situation Analysis of Equity, Diversity and Interdependence   111

PART THREE: THE WAY FORWARD

8   Reconnecting Governance and People Through Authentic Dialogue   119

9   Collaborative Planning in Action   131

Glossary   *137*
References   *147*
Index   *155*

List of Figures

1.1   Diversity, Interdependence and Authentic Dialogue Network Dynamics   6

3.1   The Geography of Northern Ireland   31

3.2   Peripheral Rural Areas in Northern Ireland   34

5.1   Sub-regional Rural Support Networks in Northern Ireland   82

5.2   LEADER 2 Local Action Groups in Northern Ireland   88

5.3   District Partnerships in Northern Ireland   92

5.4   District Councils in Northern Ireland   97

List of Tables

3.1   Population Trends in the Northern Ireland Rural Community   30

4.1   Rural Model Projects by Youth Action   49

5.1   Do You Feel Out of Place Living Here Because of Your Religion?   85

5.2   An Assessment of EDI   103

8.1   The Ten Features of Good Policy-making   120

8.2   Modes of Participation   122

The Authors

Michael Murray is a Reader in the School of Environmental Planning at Queen's University, Belfast, Northern Ireland from where he received his PhD. His research interests include partnership governance, strategic planning and community-led rural development on all of which he has published widely. His research activity takes him regularly to the United States where he has been a Visiting Scholar at Colorado State University engaged in analyses of federal-state partnership activities and community regeneration. He is the author of *The Politics and Pragmatism of Urban Containment* (1991, Avebury), co-author of *Revitalizing Rural America - A Perspective on Collaboration and Community* (1996, John Wiley & Sons) and *Partnership Governance in Northern Ireland* (1998, Oak Tree Press), and co-editor of *Rural Development in Ireland* (1993, Avebury), *Rural Planning and Development in Northern Ireland* (2003, Institute of Public Administration, Dublin) and *Participatory Governance: Planning, Conflict Mediation and Public Decision-making in Civil Society* (2004, Ashgate).

Brendan Murtagh is a Reader in the School of Environmental Planning at Queen's University, Belfast, Northern Ireland from where he received his PhD. His research interests include planning and ethnic division, urban regeneration and community development. He is a member of the Best Practice Panel of the British Urban Regeneration Association and has undertaken policy development and evaluation assignments on urban renewal in Northern Ireland. His recent book titled *The Politics of Territory* (2002, Palgrave) analyses the relationship between territoriality and land use planning. He has published widely on community-led neighbourhood regeneration.

Preface

The background to the book

The drive to tackle social exclusion and to promote a greater appreciation for diversity in the United Kingdom and Ireland are at the heart of public policy. While the emphasis of Government action has traditionally been on the cities and larger towns, attention has increasingly turned during the past decade to the problems of rural areas. Considerable effort has gone into the nurturing of an active rural citizenry as evidenced by the sheer number of community and voluntary groups participating in a wide range of development programmes. This work at a local level has been supported by an extensive infrastructure of service organisations. Their work has collectively recognised that social exclusion is more than poverty, since it encompasses all the aspects of living which enable people to be fully involved in society.

This commitment to change is being taken forward in rural Northern Ireland, where, in the context of a society fundamentally divided by politics and religion, it has long been the prevailing view that local groups are often dominated by one tradition which can result in little contact with similar groups from the "other" tradition who may be engaged in comparable activities. This lack of contact, linked to a failure to explore difficult questions and different realities, is regarded as a limiting factor to rural revitalisation. But in addition, it is accepted that the diversity of personal relationships which exist within rural service organisations can have implications for the way that an organisation is perceived internally and externally. The key proposition in this book is that if local communities are being encouraged to more fully embrace ideas of inclusiveness and diversity, then so also do the service organisations with which they interact. Moreover, while legislation can ensure minimum standards of practice and behaviour in these matters, minimum standards also tend to preclude best results from being achieved. Organisations, therefore, that progress beyond the legal imperatives and over time, through processes of authentic dialogue, remove barriers to inclusion and put mutual respect into all their relationships will contribute to societal transformation.

This book provides a baseline assessment of service organisation attitudes, behaviour and perception in relation to these issues of equity, diversity and interdependence (EDI). The narratives deal with the potential of collaborative discourse in situations where planners and public policy managers must contend with what has been dubbed "the dark side of difference". While the divided society of Northern Ireland in general, and its rural governance arena in particular, constitutes the empirical laboratory for the analysis, the aim of the book more broadly is to demonstrate not only the need for organisational change in other settings, but also to highlight possible pathways for internal and external relational transformation based on experience thus far.

Research approach

The primary research on which this book is based was commissioned from staff at the School of Environmental Planning, Queen's University, Belfast in December 2000 by Rural Community Network under the auspices of the International Fund for Ireland 'Community Bridges Programme'. The latter body seeks to support organisations which promote greater dialogue and understanding in order to tackle issues of division between people from different cultural and religious traditions in Ireland. Priority within the programme is given to four areas: community based initiatives seeking to address issues of conflict and division, especially in interface areas; intermediary groups or non-governmental organisations whose main task is to develop the capacity of communities or organisations to deal with issues of conflict; networks which may combine to cover particular geographic locations or concentrate on specific policy themes; and developing community relations dimensions and policies within institutions or across entire sectors of society. Rural Community Network is a membership organisation which seeks to provide a voice for rural people in Northern Ireland and thus the invitation extended to it to become involved in this programme is evidence of its compliance with the thrust of these criteria.

The key component of the research project has required the gathering of data through semi-structured interviews from key informants belonging to a range of service organisations with an interest in diversity and inclusion within rural areas of Northern Ireland. In line with the middle-down perspective of the research brief, it was decided to omit senior public sector officials and elected representatives from the panel of interviewees. Two interview phases were completed. In the first phase participants were selected, with advice from Rural Community Network, from a number of government related and representative/membership organisations along with key stakeholders belonging to the community and voluntary sectors. An initial letter from Rural Community Network inviting their participation in the research project was followed up with telephone calls to confirm their willingness to be interviewed. No refusals were received. A detailed topic checklist was then forwarded to facilitate preparation work by respondents. Where possible, interviews were recorded on audio tapes with the permission of interviewees. The data presented below thus directly reflect the views of those interviewed although, where sensitive comments could possibly identify respondents or their rural constituencies, we, the authors, have edited as appropriate. Some additional secondary material is included in a number of instances to enhance context appreciation.

The second phase of research interviews was carried out in the wake of the outbreak of Foot and Mouth Disease across the United Kingdom and thus required an adjustment to our methodology. Initially it was intended to convene 4 cluster conversations for participants from District Councils, District Partnerships, LEADER 2 Local Action Groups, and Sub Regional Rural Support Networks (see Glossary). This did not prove possible because of Government restrictions at that time on meetings in rural areas. Accordingly, it was decided to carry out a further round of 16 in-depth, face-to-face semi structured interviews with representatives of these area-based bodies. This period of inquiry has, without question, generated insights for the research project, particularly in relation to sensitivities connected with programme implementation, which would not have been obtained through group discussion. The narrative structure, however, seeks to retain the coherence of this group perspective by dealing with each organisational type in sequence.

A key element in the research approach has required a review of how Rural Community Network as an organisation is currently addressing equity, diversity and interdependence. Data relating to the process of internal assessment has been obtained and in this context the results of a member and non-member perception questionnaire of Rural Community Network are reported below. While it was felt to be inappropriate that Rural Community Network should seek to canvas perceptions from rural people about other specific service organisations, the view was taken that data of this type are a significant baseline measure. The approach can thus be replicated by other organisations which wish to understand more fully their specific relational position.

Finally, the opportunity was taken to convene a research seminar at which those with experience of, or interest in diversity and inclusion, could respond to the preliminary study findings. This was held in September 2001 and uniquely involved participants attending a morning session in a perceived predominantly Catholic rural community, followed by an afternoon session in a perceived predominantly Protestant community. The discussions provided an important 'reality check' for the research findings.

Reflections on the research approach

Conducting a research project on equity, diversity and interdependence produces a number of theoretical and methodological challenges. Given the highly qualitative nature of the research design, much emphasis can be placed on the value base and interpretative approach of the interviewer and this in turn raises issues about the neutrality of the research process. One strand of social science argues that applied research can never be detached and value free and that we all bring our own memories, biases and experience into the field and crucially to the way in which we frame research questions and attempt to answer them. Our reflections on the research approach adopted for this study have led us to conclude that we feel less constrained by our gender, age, social class or religious identities, but more influenced by our background as spatial planners. Our personal constructions of rurality, how we read rural society, and our shared beliefs and values about the normative agenda for rural development have had a greater impact on the carrying out of this research than other aspects of our identities and belief systems. This does not imply that experiences shaped by our gender, religious or social class are meaningless. Rather, the important point here is that the research process may, on reflection, have its origins in the distinct analytical and primarily techno-rational code that strongly underpins planning as a discipline and a profession.

We have attempted to allow for or control this process in a number of ways. Firstly, we have explicitly acknowledged from the outset that the potential exists for these values to influence our work. Second, we have employed a number of counterfactual tests in evaluating the primary data. This has included the use of quantitative data, where available, to add support to or challenge some of the qualitative commentary, the critical input by a research Steering Group and invited seminar participants into our work, and the sourcing of our arguments in the social exclusion and diversity literature, especially as it relates to Northern Ireland. However, the main way we believe that this research can guarantee at least some degree of independence is by openly stating our research evidence and presenting it in such a way that it can be subjected to multiple and contradictory meanings. Accordingly, we make no apology for explicitly presenting the data in the form of quotations, many of which are deliberately long, in order to aid that re-interpretation. Our conclusions can be more fully appreciated in the context of this raw data.

Equity, diversity and interdependence are interlocking elements of a language informing our work and are described throughout this book as a journey of discovery, rather than as an object or a product. Our personal journey in the interpretation of rural life through the lenses of EDI, our need to build confidence with interviewees and our desire to establish shared understanding with our professional colleagues are all reflected in the narratives below. We believe that conversations on the matters included in this book are necessary within all organisations located within the rural development arena. It is our view that the research process contributed to the opening up of that space for dialogue. Thus we have attempted to report our interviews faithfully and to record the weight of opinion and priorities raised. However, while this may appear at times to offer a varying emphasis on issues, for example, in relation to young people, it should be regarded more as an indication of perception breadth and depth across territories, sectors and interest groups.

Book structure

The book is divided into three parts. In Part One the conceptual and practice background of the research is reviewed. Chapter 1 locates our research in the literature on ethnicity, race and place and, in particular, highlights the relevance of discourses on equity, diversity and interdependence for spatial planning outcomes and planning practitioners. Chapter 2 focuses more specifically on social exclusion, social prejudice and sectarianism in Northern Ireland. Having reviewed the sweep of official discourses on these inter-related matters, the chapter goes on to introduce the alternative action framework of equity, diversity and interdependence as an approach through which to secure greater inclusiveness within the relational setting of organisations.

Part Two concentrates specifically on the research laboratory of rural Northern Ireland and commences in Chapter 3 with a short overview of governance arrangements within that arena. This is followed by a comprehensive reporting of the qualitative and quantitative data gathered during the interviews and questionnaire survey. The content deliberately reflects the perceptions of respondents rather than the interpretation of the authors of this book and thus extensive use is made of quotations derived from interview tapes. The narrative is structured on a thematic basis in Chapter 4 and around area-based organisations in Chapter 5. Chapter 6 reviews the progress being made by Rural Community Network to more fully deal with issues of diversity and inclusion at an organisational level. This case study of the internal experience, thus far, is offered for the reason that external credibility is needed before any organisation can seek to give support to others wishing to participate in comparable processes. Chapter 7 draws together the data collected during the research project and presents a situation analysis of perceptions, behaviours and challenges associated with equity, diversity and interdependence. This synthesis provides a necessary linkage, therefore, between the previous 'report of survey' and insights offered on the way forward in Part Three of the book.

In the final section Chapter 8 revisits a number of important debates dealing with the relationship between governance, social inclusion, social capital formation and authentic dialogue. It is against this backcloth that the contribution of EDI to broader social cohesion is explored, not least within societies seeking to move beyond mistrust, inequality, hatred and violence. In the concluding Chapter 9 the specific implications of EDI for collaborative spatial planning are assessed. The key point here is, that in situations where place is contested, it is possible to expose the nature of these contexts, identify where alliances are possible, and think about how fragile connections might be supported. Authentic dialogue and trust are essential in moving forward a prescriptive agenda for change.

Acknowledgements

The completion of this book has depended upon the generosity of many people in terms of their time and insights. We wish to thank all those from Northern Ireland-wide and area-based rural service organisations who met with us during the project and who, through the honesty and quality of their comments, provided us with the depth necessary for this baseline analysis of equity, diversity and interdependence. The contribution of all participants at a full day research seminar is gratefully appreciated. We also extend our gratitude to those individuals who completed the Rural Community Network perception questionnaire.

Within Rural Community Network we received strong support for this research project from Tony Macaulay, Niall Fitzduff, Roger O'Sullivan and Michael Hughes. The Steering Committee appointed by Rural Community Network provided valuable feedback on draft position papers and in this regard we especially express our appreciation to Karin Eyben from University of Ulster for her valuable critical reflections on our work.

We are pleased to acknowledge the financial assistance of the International Fund for Ireland by way of a research grant to Rural Community Network. We extend our appreciation to Joe Hinds in his role as Co-ordinator of the Community Bridges Programme for his enthusiasm and encouragement.

We are delighted that Ashgate contracted to publish our research within the series 'Perspectives on Rural Policy and Planning'. In this regard we would like to thank Valerie Rose and Carolyn Court for their assistance and no small amount of patience in the run up to publication. The comments received from an anonymous reviewer have been especially helpful in finalising the manuscript.

Thus, while many individuals have contributed to this publication, it is important to state that what follows is the work of the authors of this study who bear joint responsibility for any unintended inaccuracies. We hope that our research can contribute not only to an ongoing authentic dialogue in situations where there is division, but also to a real search for understanding and mutual respect.

Michael Murray and Brendan Murtagh,
Queen's University, Belfast.

PART ONE:
SETTING THE CONTEXT

Chapter 1

Discourse, Planning and Place

Introduction: the dark side of difference

As regions are increasingly shaped by the global flow of knowledge and resources, the notion of 'place dis-embeddedness' has held important attractions for planning theorists. The restless search to steer advanced capitalist states through this uncertainty has focused particular attention on new forms of flexible governance for reconciling multiple and conflicting interests and on a concern for knowledge in working through policy outcomes. 'Collaborative planning' identifies decision-making structures, styles and routines as critical components in plan formulation and exhorts planners to guarantee stakeholders access to policy arenas concerned with the quality of places (Healey, 1996). The normative task is mediation and filtration through the sets of conflicting interests to produce some form of consensual agreement about the use and development of land. This concern for discursive analysis, with its roots in Habermasian communicative action, has emerged as the dominant paradigm in planning theory (Tewdwr-Jones and Allmendinger, 1998). Moreover, its reach has been felt in planning practice through renewed interest in public participation, multi-sectoral area based partnerships and visioning processes within rural and urban communities.

This book deals with the potential of collaborative discourse through authentic dialogue where planners and, more generally, public policy managers must contend with "the dark side of difference" (Sandercock, 2000, p.14). The claims, experiences and entitlements of place bounded ethnic or religious communities are often not reducible to the sorts of neo-liberal management processes assumed by collaborative planning. Deeply divided and fabricated communities, formed as "sites of resistance" (Parker, 2001, p.195), are defined by their distance from, not proximity to, official discourses about planning, regeneration policy or area-based regeneration structures.

Accordingly, the chapter begins by briefly reviewing the contribution of collaborative planning to understanding divided societies. We suggest that authentic dialogue could help redefine the role of planning in dealing with both *diverse* and *interdependent* environments. The discussion then highlights a suite of professional responses to issues of race, ethnicity and poverty. We demonstrate that Equity, Diversity and Interdependence (EDI) offer a framework to advance the notion of discourse in divided societies where 'place' conditions the nature of the relationship between competing interests.

Collaborative planning and EDI

In his work, Habermas (1987) identifies the importance of 'life worlds' where personal experiences and the routine of daily life have the power to shape both market and political systems. The imposition of technical-rational discourses in our life worlds has crowded out moral and emotional concerns and the way in which people value and prioritise their every day experiences. Communicative theorists argue that different modes of reasoning and systems of meaning have equivalent status in debate and that the task for planners is to develop strategies for collective action through interaction and dialogue (Healey, 1997, p.53). Here, language is vital and in planning the priority is to establish a process of interactive collective reasoning or discourse which, in turn, involves a degree of collaboration, trust and reciprocity (Habermas, 1984). "In the end, what we take to be true and right will lie in the power of the better argument articulated in specific socio-cultural contexts" (Healey, 1997, p.54).

The *institutional approach* recognises that human actions and discourses are played out within the context of broader economic, labour market and political structures. But, unlike Marxist analysis, it suggests that individuals are not passive receptors of the working out of these systems but are reflexive agents with the choices and capacity to modify and even transform the structuring forces that influence their life worlds. Institutional theorists emphasise the importance of social relations and interaction with others in making sense of communities and places and this process of creating and maintaining relational bonds has clear connections with scholarly and empirical interest in social capital, not least within a Northern Ireland context.

For Taylor (1998) social capital lies at the heart of local power circuits whose sophistication will dictate the radicalisation of community politics and the effectiveness of neighbourhood partnerships. Putnam defines social capital as the "features of social organisation, such as trust, norms and networks that can improve the efficiency of society by facilitating co-ordinated actions" (Putnam, 1993, p.167). Communities that have high levels of trust, networks and organisational capital are better informed, better led and more content than those which do not. "Moreover, when individuals belong to cross-cutting groups with diverse goals and members, their attitudes will tend to moderate as a result of group interaction and cross pressures" (Putnam, 1993, p.90). But the way in which social capital is assembled and reproduced is path-dependent and history can reinforce social pathologies, patterns and institutions, even if they are inefficient. Thus, for example, the development of social capital in Northern Ireland is intimately wedded to expressions of national identity and to asymmetrical efforts to survive, protect and grow. In highly segregated areas this process of institution building, maintaining community capacity and defining rigid territorial boundaries has constrained the possibilities for the trust which is vital for wider strategic engagement. This experience resonates with the wider observation by Putnam that "dense but segregated horizontal networks sustain co-operation within each group, but networks of civic engagement that cut across social cleavages nourish wider co-operation" (Putnam, 1993, p.175).

Writing in a Northern Ireland context, Porter (1998) has argued that exclusive, zero-sum politics provides little space for mutual recognition, respect or dialogue. Like Putnam it is in the sphere of civil society that Porter sees the best hope for the future:

> If citizenship is about participation and not just about representation and if politics is creative and not merely proactive, then occupying the public space that civil society makes available is the best antidote to citizenship alienation from the bureaucratic state and growing political apathy. (Porter, 1998, p.202)

Civil society, notwithstanding its multiple meanings, is where voluntary organisations, community groups and individuals make claims, assert rights and enjoy entitlements independently of politics. It provides an alternative arena for deliberating issues of social justice, separately from political organisations and autonomous from state control. In short, Porter argues that the search for political stability requires inclusive public places open to disparate voices in society:

> It is by arguing with and listening to one another, by proposing and counter-proposing, agreeing and disagreeing, evaluating and judging together, that we may hope to reach decisions about our collective life which we can all be committed to. (Porter, 1998, p.211)

Healey has proposed a methodology for implementing and evaluating the potential of such an agenda empirically. Her *Institutional Audit*, unpacks the circumstances, settings and routines through which various stakeholders have a say in policy making and offers a framework to explore further the political subtleties embedded in resource allocation decisions. This involves asking a range of inter-linked questions:

- Who has a *stake* in the qualities of places; how far are these stakeholders actively represented in current governance arrangements?
- In what *arenas* does discussion currently take place? Who gets access to these? Do they interrelate issues from the point of view of everyday life and the business world? Or do they compartmentalise them for the convenience of policy suppliers?
- Through what *routines* and in what *styles* does discussion take place? Do they make room for diverse ways of knowing and ways of valuing representation among stakeholders or do dominant styles dominate?
- Through what *policy discourses* are problems identified, claims for policy attention prioritised, and information and new ideas filtered? Do these recognise the diversity among stakeholders?
- How is *agreement* reached, how are such agreements expressed in terms of commitments and how is agreement monitored? Is it easy for those who are critical to implementation of the agreement to escape from the commitments? (Healey, 1996, pp.213-4)

The point here is that the connection between discursive planning, social capital and Equity, Diversity and Interdependence has important theoretical and empirical appeal. Booher and Innes (2002) argue that the informational age, the break-down of traditional economic power in primary industries and the loosening of the class structure has emphasised what they term 'Network Power' which "emerges from communication and collaboration among individuals, public and private agencies, and businesses in a society. Network power emerges as diverse participants in a network focus on a common task and develop shared meanings and common heuristics that guide their action" (Booher and Innes, 2002, p.225). Success will be guaranteed, they suggest, when *Diversity, Interdependence and Authentic Dialogue (DIAD)* are present as shown in Figure 1.1.

Figure 1.1 Diversity, Interdependence and Authentic Dialogue Network Dynamics

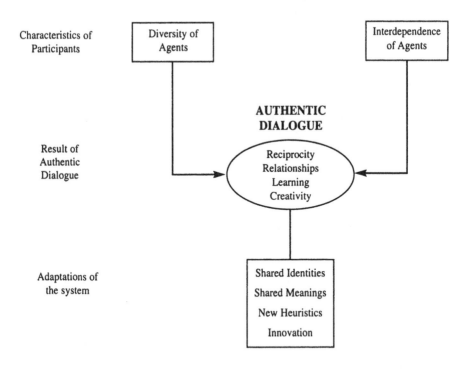

Source: Booher and Innes, 2002, p.227.

We contend that this provides the substantive theoretical or conceptual lenses with which to explore EDI in the context of societal divisions. Booher and Innes frame it this way:

> Interdependence among the participants is the source of energy as it brings agents together and holds them in this system. Authentic dialogue is the genetic code, providing structure within which agents can process their diversity and interdependence ... Diversity is the hallmark of the informational age. The wide range of life experiences, interests, values, knowledge and resources in society is a challenge for planning and the efforts to produce agreements and collective action. (Booher and Innes, 2002, p.227)

For planning to be mobilised as a dialogical process it needs a new methodology in which information "influences by becoming embedded in understandings, practices and institutions, rather than being used as evidence" (Innes, 1998, p.52). This concept of information and research provides the basis of the methodological perspective adopted here. However, the central criticism of communicative planning is its approach to power relations. Thus the DIAD model makes little reference to how diversity and interdependence relate to inequity within society and the sets of interests that are encouraged to discourse with one another. Yiftachel, for instance, makes the point that collaborative planning focuses on a 'critical commentary about planning' rather than a 'societal critique of planning'. The emphasis should not concentrate on the conduct of planners and their practices but rather on the broader power structures and "legitimisation dynamics within which public agencies often act" (Yiftachel, 2001, p.253). Similarly, Tewdr-Jones and Allmendinger (1998) point out that different interests have different access to information and can mobilise and interpret knowledge in vastly differing ways, especially within the planning system. It is our view that equity must be intimately connected to diversity and interdependence in order to build the sort of radical agenda required to address deepening social, ethnic and spatial tensions in divided societies.

Thus, for Soja, it is space that represents the crucial category for understanding the connections between all sorts of multiple oppressions and the possibility to mobilise around agendas framed by place. The social production of space is "an important strategic milieu for a new coalition politics of class, race, gender, sexuality, age, ethnicity, locality, community, environment, region and other sites and sources of cultural identity and the assertion of difference" (Soja, 1999, p. 75). Sites of resistance have emerged where economic and racial inequality have interlocked to produce fractal places characterised by spatial concentrations of hopelessness and alienation (Parker, 2001; Soja, 2000). Heikkila (2001) argues that this conception of space represents the key challenge to land use planners:

> Space matters because it mediates the experiences of people in places, and further, it shapes the structure of the opportunity set available to them. When this transpires to create a stagnant population pool of poverty cut off from the mainstream, the poverty deepens and becomes more profound in character and more potent in its ability to sustain itself over time. (Heikkila, 2001, p.266)

Similarly, Qadeer (1997) has argued that the multi-nationalism of the global economy is further diversifying built forms and functions. This calls for a distinct response within the policy, procedural and value base of the planning profession:

> A fundamental effect of multiculturalism is to call for pluralistic planning approaches and to question unitary conceptions of public interest and the ideology of master plans. (Qadeer, 1990, p.482)

Healey (1999) has specifically recognised the potential of multi-cultural encounters to be submerged by the dominance of the more powerful status group, the powerlessness of weaker groups or the endless fracture that often characterises ethnic disputes. On a more positive note she observes that:

> governance practices that seek to sustain the diversity of relational webs in a locality may create processes and arenas with the potential to generate multifaceted learning, build intellectual and social capital and through developing shared systems of meanings create a new 'relational world', with a new layer of cultural formation to add to the store already available in a locality". (Healey, 1999, p.116)

Nevertheless the text of the discourse is often violent and hegemonic. We show later in our empirical analysis of rural Northern Ireland that the discursive relational webs are themselves embedded in exclusive ethnic ideology, beliefs and organisations. What is needed is a deeper understanding of the way in which space is regulated by ethnicity and of how the combination of ethnicity and religion conditions the possibility for meaningful dialogue. Considerations of diversity, interdependence and equity are crucial to that understanding.

Discourses on diversity

Thomas argues that a professional ideology about the technical content of planning, the needs of primary industries and property, and powerful lobby groups, especially in middle class areas, has left little room for the discussion of race in a British context. "However, in the 1970s and 1980s the needs of black and ethnic minority residents forced them into greater contact with the planning system; in addition, politicians and community activists in some areas became more aware of the existence and significance of planning" (Thomas, 2000, p.77). A number of global, issue based, professional and more localised or regional factors have driven a shift in consciousness and practice within planning and its attempts to wrestle with racialised places.

Thus Fenster (1996) has explored the use of planning powers and processes to control Arab land rights and entitlements. She argues that discrimination could arise when ethnic groups are treated differently from the majority or when minorities are treated similarly in different situations. The former denies citizenship rights whilst the latter denies the uniqueness of ethnic groups (Fenster, 1996, p.415). Yiftachel

(2000) proposes that planners have a choice of three roles when confronted with these sorts of dilemmas: in an assimilation model planners attempt to dissolve difference based on the primacy of citizenship rights; a pluralist model recognises distinctive ethnic identities and responds diversely to the circumstances and needs of different ethnic groups; and thirdly, a discriminatory stance enables the planner to manipulate space to subjugate minority or powerless interests (Fenster is careful to locate Israeli planning history in this context). But these alternatives are not confined to highly politicised or territorial conflicts. For example, the UK government Home Office response to race riots in the northern British cities of Bradford, Burnley and Oldham in 2000 can be read, albeit loosely, within an assimilationist model. Here, bipolar politics and identities were the targets for a renewed initiative on community cohesion with a fairly selective rendition of British citizenship underlying that policy drive:

> A civic identity, which serves to unite people and can express common goals and aspirations of the whole community, can have a powerful effect in shaping attitudes and behaviour. Shared values are essential to give people a common sense of belonging regardless of their race, cultural traditions or faiths. (Home Office, 2001, p.12)

In South Africa, the project of residential integration has transformed the policy, if not the urban landscape of the new nation. Turok (1994) has traced the transition from apartheid to democracy and, in particular, the effort invested in strategic planning to deal comprehensively with poor township infrastructure, to diversify buffer zones previously used for isolating black areas, and to construct appropriate forums for including community interests in the planning process. However, Christopher (1998) points out that changes to the Group Areas Act have been slow to emerge because of the weak economic position of the majority black population and because the minority white population has remained in cities as a significant political, economic and social bloc. Whilst there have "been no land invasions" (Christopher, 1998, p.234), specific planning measures have been used to break up rigid residential segregation including the use of buffer zones for low income housing, redevelopment schemes in the inner-city and community development programmes. Bollens (2002) has pointed out that market based normalisation in South Africa has created new class cleavages spatially manifest in the movement of jobs and people to high-income suburbs. Seekings (2000) reports that there have been only limited schemes to undo the apartheid city, new social housing provision is limited and the quality of public services comparatively poor. It is against that backcloth that   Ramutsindela (2001) identifies the central tension between the process of de-radicalisation and nation building and the responses to the new aspirations of marginalized black people.

Similar dilemmas have confronted British state management of the Northern Ireland conflict. A deep and violent crisis was met with a multiple response involving militarisation, social reform and legislative guarantees for the Catholic minority, especially in the workplace. An explicit assimilationist project was supported with the establishment of the Community Relations Council, the promotion of integrated education and an attempt to redress imbalances in the economic conditions of Catholics and Protestants (Shirlow, 2001). The formation and maintenance of a loose form of material citizenship offered an opportunity of breaking down atavistic identity constructions. Economic restructuring and equality of opportunity guaranteed new mobility for a growing Catholic middle class who could identify the benefits of apolitical lifestyles and aspirations (Murtagh, 2002). Planning played an important role in modernising the region, providing new consumption opportunities for rising disposable income and projecting positive place imagery to global investors and tourists. This class realignment was reflected in tenure restructuring with a residualised public sector housing stock, containing comparatively even numbers of Catholics and Protestants and new suburban as well as gentrified housing spaces responding to socio-economic mobility.

Within the context of diversity, therefore, what seems to be happening is that new spatial cleavages are overlain on traditional territorial landscapes. But in a region such as Northern Ireland old enmities are most viciously contested where urban and rural communities feel abandoned by economic change and largely untouched by progressive social need or equity policy. Assimilation is checked by plurality and its deeply socio-political and spatial character. The modernising (or post modernising) of the region has left little space for any serious consideration of the planning specifics of territoriality and residential segregation. The important point here is that the formulation and implementation of planning policy has failed to meaningfully respond to the spatial effects of segregation, territoriality and social distance. Indeed Bollens (1999) has pointed out that the 'colour blind' policy adopted in Northern Ireland does not leave planning as an unproblematic variable producing and reproducing inequality or new geographies of opportunity. He is critical of the assumptions made by neo-liberal collaborative scholars that planning is a force for good but needs to be better networked and popularised. The evidence that planners have produced consensual or collaborative responses to ethnic-spatial division is limited and slow to emerge. A successful working through of the collaborative model might at least expose the planning system to the viability of alternatives to segregation or even the desirability of integrated living. Our argument is that EDI provides a framework to enable this necessary transformative agenda within organisational cultures and policy routines.

Discourses on interdependence

Marc Seitles (1996) was especially critical of the role of public policy in producing residential segregation in American cities:

> The devastating effects of residential racial discrimination on the quality of life for minority families and for our culture at large, represent the importance of initiating policies to integrate residential neighbourhoods. Without the efforts of integration, the negative effects of decades of bigoted housing policies will be exacerbated, therefore perpetuating the existence of segregation and racial division. (Seitles, 1996, p.17)

Seitles recommended a radical programme that included inclusionary zoning techniques, which would ensure that a range of low-income housing opportunities could be available within private sector developments. Mobility programmes have helped to support the relocation of people to new areas and housing opportunities. In Chicago, for example, a scheme implemented by the Chicago Housing Authority (CHA) achieved the relocation of 5,000 families by 1993; some 84 per cent of those who moved to non-concentrated areas felt their quality of life had improved. In addition, the CHA is under mandate to construct public housing in predominantly white neighbourhoods and by 1993, 591 units had been provided. The Cincinnati Metropolitan Housing Authority also housed more than 600 families using the mobility scheme between 1984 and 1993. As a result, black suburban residents reported a 57 per cent employment rate, compared with a 24 per cent rate among those still living in public housing. Families also reported that they did not experience racist behaviour from their new neighbours, liked the schools better than those in the inner-city and worked in higher paid jobs with increased benefits (Seitles, 1996).

Dorsett (1998) points out that the economic marginalisation of ethnic minorities means they are often forced to accept worst housing conditions and that discrimination reduces their residential choices. He has distinguished between strategic reasons for segregation, such as, location relative to suitable labour markets and tactical reasons, such as, proximity to friends and families or places of worship. Both processes need to be understood in the analysis of segregation and assimilation. Planning policy must achieve a delicate balance and reflect the fact that both choices and constraints are important in determining residential location. While the deprivation of those areas in which ethnic minorities live is a clear-cut issue deserving attention, tackling segregation is not so straightforward – many individuals from ethnic minorities will choose to live in areas where their own ethnic group or other minority groups, are well represented (Dorsett, 1998, p.x). Again the framework of EDI can help identify and evaluate the possibility for group interdependence and what this might mean for shaping local communities through spatial planning interventions.

Discourses on equity

In an exhaustive evaluation of UK policy on race and residence, Smith (1989) concluded that lack of access to public sector housing and the concentration of ethnic minorities in certain estates built up a problem that the Urban Programme proved ineffective in addressing. Specifically, the resources devoted to urban policy throughout the 1980s were comparatively small, the policy targeted toward areas missed many vulnerable groups and individuals and cuts in welfare expenditure affected black people adversely. These criticisms were also levelled at the Single Regeneration Budget (SRB). In *Race and Regeneration*, the Local Government Information Unit (LGIU, 1995) found that the majority of successful bids did not prioritise ethnic minority issues, there were no ethnic minority-led successful bids and there were few Section 11 (Race Relations) type SRB Bids funded (LGIU, 1995, p.8). In their evaluation, Mawson *et al.* concluded that "There is a strong case for considering that a proportion of the SRB be temporally ringed fenced to finance ethnic minority and S11 type projects. This is required for two purposes: 1. to build the capacity of ethnic minority organisations; 2. to reinforce the importance of ethnic minority issues within regeneration strategies" (Mawson *et al.*, 1995, p.137). The Bidding Guidance to the Fourth Round of the SRB was subsequently amended to ensure that it could "complement other work to tackle racial violence and harassment in local communities, and also target economic development and training initiatives on such communities" (DETR, 1997, p.3).

Oc *et al.* (1997) have examined the experiences of ethnic minorities in training and business support in City Challenge areas. Their research echoes many of the previous criticisms of SRB implementation. Because the issue of ethnicity was not prioritised in the guidance, it was left to individual authorities to design the needs of marginal groups into their programmes. Oc *et al.* located their conclusions within the overall need for the planning system to recognise the distinctive needs of ethnic minorities in order to enable them to compete better in the mainstream economy. They suggested that in areas of "ethnic minority concentration (i.e. in excess of 10 percent of the population), there should be a comprehensive needs analysis to enable effective programme design and monitoring" (Oc *et al.*, 1997, p.viii). They highlighted the need to ensure that ethnic minorities were represented on local management structures in order to develop a sense of local ownership of regeneration programmes and training schemes.

This issue of representation in urban regeneration governance structures was also raised by Brownhill *et al.* (1996). They pointed out a contradiction in Urban Development Corporations whereby the issue of race itself had been marginalized, but local structures afforded the opportunity for minorities to exercise control over decision-making and resource allocation. However, the degree to which in practice they could exercise that control was variable and dependent on local circumstances, personalities and the strength of community networks.

A theme running through this analysis of race, equity and planning is the need to monitor the impact of policy on minorities and develop new ways of involving them in the planning system. Riley (1994) has described the use of 'Race Equality Progress Reports' in Sheffield to evaluate ethnic impacts on an annual basis. These reports aimed to act as an early warning system where there was a high probability of refusal of applications from minority groups. These applications would then receive greater attention from planning staff. Ahmed and Booth (1994) have described the appointment and impact of a race advisor to a planning department. The officer had three functions:

1. To advise planning officers and elected members on racial issues and matters which concern ethnic minorities;
2. To raise awareness and sensitivity of staff to racial issues; and
3. To raise the awareness and encourage the involvement of the ethnic minorities in the planning process.

An 'issues document' was prepared, for example, as part of a Central Area Local Plan which attempted to highlight the main issues of concern to ethnic minorities. This was ultimately translated into guidelines on the needs of ethnic minorities and was underpinned by Racism Awareness Training for all staff in that planning department. The guidelines then formed part of the basis for drawing up policies for the draft Unitary Development Plan.

Substantive concerns about ethnicity, equity and planning have also been reflected in debates about professional competency and responses to multiculturalism. In Britain, the Royal Town Planning Institute (RTPI) and the Commission for Racial Equality (CRE) established a Working Party to examine linkages between race and planning. It suggested the need for greater communication between minorities and planners, the introduction of translation services and the need for sensitivity when dealing with applications with an ethnic dimension such as places of worship or eating establishments. The need for monitoring the impact of policies on ethnic groups, for race relations training aimed at planners, and for a distinctive consultation process with minorities were also stressed (RTPI/CRE, 1983). The array of proposals thus represented an impressive recognition of the ethno-spatial context of planning and a shift in ideology away from its technocratic roots. Many of these proposals were restated a decade later in an RTPI study on Ethnic Minorities and the Planning System (Krishnarayan and Thomas, 1993) which identified three weaknesses in current procedures. These were the relatively low priority given by planning authorities to service delivery to minorities, the uncertainty about the extent to which issues of race could be taken on board in planning decisions, and the underdevelopment of good practice guidance. The study emphasised that race sensitivity should not be considered a side issue or luxury but should permeate the culture of the profession. A code of practice for planning was suggested, as was strengthening the role of the Race Relations Act in development plan and development control decisions. Thomas revisited the study in 1997 and highlighted the need for planners to be aware of their role in reproducing insensitive policies and to understand the experiences of ethnicity as a basis of plan formulation and implementation (Thomas, 1997, p.209).

This analysis has resonance in contemporary debates about racism and planning in the United States. Hoch (1993) has argued that understanding minority experiences of deprivation and exclusion is a starting point to reorienting the profession and sensitising it to issues of race and injustice. Manning-Thomas (1994) has argued that direct and indirect planning decisions marginalised and concentrated ethnic minorities. Inner-city redevelopment, the concentration of industry in mainly white areas, suburban ordinances restricting choice for low income families and transport systems designed for commuters had prolonged negative effects on the black urban experience. Following this, Sandercock (1998) set out an agenda for planning professionals premised on the needs of a multicultural society. At the heart of this was a move away from technical rationality in planning methodology and towards ethnographic techniques that record the experiences of a range of marginal groups, effective listening strategies as a basis for development planning and a respect for ethnic diversity and how it is spatially expressed. This is precisely the objective of an EDI approach to organisational change which we argue is especially relevant in a Northern Ireland context where planning, equity and social need have been technocrised and systemised over 30 years of conflict. Our analysis and good practice insights have, however, a much wider application as evidenced by the depth and breadth of challenges rehearsed in this introductory chapter.

Conclusion

Discursive planning is at the forefront of planning theory, yet the evidence on where and, crucially, how it is delivered in practice has commanded less space in the literature. More importantly, the potential of *discourse* in places stratified by violence, fear, historical mistrust and inequality needs to be rigorously evaluated if planning is to claim any relevance in marginal communities, whether in urban or rural regions, and in either advanced or transitional capitalist societies. The fine-grain of discourse is unpacked in this book, especially in the way it shapes the perspectives of interests that are dialogically separate and antipathetic to one another. Our laboratory is rural Northern Ireland, but our argument is that a discourse framed by the principles of equity, diversity and interdependence has the potential to realign multiple interests in more meaningful and purposeful ways in the context of an uncertain progression from violence and inequality to political stability and social cohesion.

Chapter 2

Introducing Equity, Diversity and Interdependence

Introduction

In March 2002 the Equality Commission for Northern Ireland published a consultation document titled *Good Practice Guide to Promote Racial Equality in Planning for Travellers*. The need for this advice arises from the perceived discrimination against Travellers in the statutory system of town and country planning. A number of recommendations are advanced by the authors of the report (Ellis and McWhirter, 2002). These outline the suggested content of clear planning policies, identify the need to engage the Traveller community in meaningful consultation on planning issues, and present guidelines on how to deal with racist planning objections.

In January 2003 The General Synod of the Church of Ireland published *The Hard Gospel: Dealing Positively with Difference in the Church of Ireland*. This report was prepared by the church's Sectarianism Education Project Committee and presents a unique survey of attitudes to sectarianism and other forms of difference, not least those arising out of differences in political allegiance between members of the church in the separate jurisdictions on the island of Ireland. The analysis, however, moves beyond sectarianism to also consider peacebuilding initiatives, ethnic differences, sexuality and relationships, and age-related issues.

In June 2003 the Rural Development Council for Northern Ireland published *A Picture of Rural Change* comprising its second annual baseline report examining patterns of activity and change across rural Northern Ireland. The analysis spans a wide range of economic and population data, but includes fresh evidence relating to the spatial distribution of some key public services in rural areas, for example, Post Offices, GP surgeries, pharmacies, day-care facilities, primary schools. Other aspects related to the social wellbeing of rural people are mapped and include community halls, crime, community groups and community group activity.

These three reports, selected for illustrative purposes, are but a few of the very many which have been issued in recent years dealing with societal diversity, the complex interdependencies between governance and people, and the challenges faced by public managers in responding to imperatives of equality. Simply stated, contemporary society is stratified by multiple exclusions that interlock to deepen the isolation experienced by some people and interests. This chapter explores these realities by taking social exclusion as a key departure point for a follow-on analysis of social prejudice and sectarianism. Official discourses relating to community relations, equality and social inclusion are then examined in order to assess their contribution to the systemic change of rooted behaviours. The limitations of these perspectives on societal transformation are identified. It is at this point that we then turn to the framework offered by EDI which provides a more meaningful way to think creatively about how to move these official discourses beyond their conventional legislative and institutional boundaries. While much of the discussion is located within the parameters of policy and circumstance in Northern Ireland, the underlying concepts, argument structure and conclusions have wider relevance.

Social exclusion

The social exclusion discourse was originally coined in France in the mid 1970s and was used to categorise people unprotected by social insurance (Cousins, 1998). In the 1980s the term was used to apply to a broader process of social disintegration which involves a fracturing of the relationship between the individual and society, particularly those affected by the economic restructuring prevalent at that time. From France the phrase spread to other European countries and was adopted in the language of the European Commission in 1989 as an expression related to poverty and the inadequate realisation of social rights. Social exclusion has thus been linked at the EU scale to the barriers (and their consequences) to the social rights of citizenship, to a basic standard of living and to participation in the principal occupational and social opportunities of society. During the 1990s the phrase was widely adopted and in Great Britain was given priority status by the Labour Government in 1997 with its newly established Social Exclusion Unit. The thrust of its work is illustrated by the theme of an early report which offered a national strategy for poor neighbourhoods (*Bringing Britain together: a national strategy for neighbourhood renewal.* Cm4045, 1998). Again the focus on poverty outcomes is prevalent.

However, the phrase social exclusion is used in many ways. Shucksmith (2000, 2001) identifies three perspectives associated with its definition:

- an integrationist approach in which employment is seen as the key integrating force, both through earned income, identity and self worth, and networks of personal support;
- a poverty approach in which the causes of exclusion are related to low income and a lack of material resources, for example, adequate food, clothing, shelter;
- an underclass approach in which the excluded are regarded as deviants from the moral and cultural norms of society, exhibit a dependency culture and are blamed for their own poverty and its transmission from one generation to the next.

These have been summarised as "no work", "no money" and "no morals" respectively. Shucksmith's analysis, however, takes us further than a focus on poverty, distributional issues and "victims". It leads towards an appreciation of system failure, the processes which cause exclusion and the importance of local context in such processes. Thus the emphasis of social exclusion is on relational issues (low participation, lack of social integration, powerlessness).

These matters have been explored by Atkinson (1999). He identifies four dimensions to system support:

- the democratic and legal system, which promotes civic integration;
- the labour market, which promotes economic integration;
- the welfare system, which promotes social integration; and
- the community and family system, which promotes interpersonal integration.

Thus social exclusion occurs when one or more of these systems breaks down. Atkinson argues that social exclusion is not just, therefore, about unemployment and lack of income, but also a wide range of living conditions such as community, health, education, housing, safety etc which are embedded in the institutional systems. While breakdown or malfunctioning in any of these components can generate social exclusion, Atkinson identifies the deepest challenge for policy makers as occurring when there is multiple system breakdown or malfunctioning either simultaneously or as part of a chain reaction. For example, an individual or group poorly integrated within the community system when faced with long term unemployment may encounter social isolation which a constrained labour market, exacerbated by governance instability, will accentuate towards social exclusion.

Harvey (1994), in his examination of the lessons from the Third EU Poverty Programme in Ireland, synthesises the threads of the analysis above with a definition of social exclusion which has real meaning for this EDI research:

> Social exclusion is a much more dynamic concept of the processes of social change than poverty. Social exclusion draws attention to its underlying causes as much as its manifestations. Social exclusion refers to the structures and processes which exclude persons and groups from their full participation in society. It explains that poverty does not just happen: it flows directly from the economic policies and the choices which society makes about how resources are used and who has access to them. The forces of exclusion change as economies and societies change. Social exclusion may take a combination of forms - economic, social, cultural, legal - with multiple effects. The term exclusion has connotations of process, focusing on the forces by which particular categories of people are closed off from the rights, benefits and opportunities of modern society...Social exclusion is not just about lack of money, but may be about isolation, lack of work, lack of educational opportunities, even discrimination. The notion of social exclusion has a strong policy focus: it is often the result of the ineffectiveness of policies, of the perverse effects of policies and of the distorting outcome of social class decisions. Social integration or inclusion, by contrast, is about drawing people into society in a number of different, complementary ways - into the labour market, into social services, into more equal relationships with their fellow citizens, into networks of care, companionship and personal and moral support. (pp.3-4)

Social inclusion thus turns the concept of social exclusion on its head, by identifying the positive side of the same coin. It encapsulates the processes whereby all members of a society are enabled to participate in its social, economic and civil activities. However, many are denied that opportunity as a result of social prejudice.

Social prejudice

Social prejudice provides the necessary conceptual linkage between social exclusion and sectarianism and generates the necessary space within which to locate equity, diversity and interdependence as a response to these sets of malaise. In a comprehensive analysis of prejudice in Ireland, MacGreil (1996) defines social prejudice as:

> a hostile, rigid and negative attitude towards a person, group, collectivity or category, because of the negative qualities ascribed to the group, collectivity or category based on faulty and stereotypical information and inflexible generalisations. (p.19)

In explicating this definition MacGreil highlights the following matters:

- that while the operation of prejudice towards the group, collectivity or category is normally towards the individual, there are occasions when the prejudice may be exercised towards the group, collectivity or category at one remove from the individual, for example, towards Travellers *per se*;
- the rationalisation of prejudice through self justification tends to marginalise the person or group;
- prejudice becomes deeply rooted in culture, social structure and interpersonal response traits.

It is against these considerations that MacGreil identifies the manifestations of social prejudice as *(1) ethnocentrism* – conviction that members of other cultures or nationalities are inferior to one's own culture or nationality because of their culture, nationality or way of life; *(2) political prejudice* – rejection of people deemed to belong to certain political groups or categories, because of their political views; *(3) religious prejudice* based on perceived religious affiliation or non affiliation and encompassing sectarianism; *(4) sexism* which incorporates prejudice based on gender; *(5) homophobia* which focuses on attitudes and behaviour related to sexual orientation; and *(6) other social prejudice categories* extending from ex-prisoners, people with mental and physical disability, people who are unemployed, to unmarried mothers. The value of this classification, regardless of how the content may be debated, lies with its acknowledgement that we live in societies of complex group interactions. These interactions, MacGreil suggests, generate multiple responses and differentiated public policies. These can change relationships in a negative fashion, for example, by avoidance and withdrawal, fear and anxiety, expulsion and segregation, aggression and conflict. On the other hand interactions can be positive, say through a commitment to pluralism, whereby group differences are to be facilitated and respected, there is equality of treatment for all groups, and there may even be duplication of facilities. All these considerations resonate loudly within Northern Ireland where, traditionally, the prejudice debate has long been dominated by sectarianism.

Sectarianism

Sectarianism in Northern Ireland has generated a considerable literature (for example, Liechty, 1993; McMaster, 1993; Connolly, 1998; Miller, 1998; Bryan, 2000; Liechty and Clegg, 2001; Higgins and Brewer, 2003) and thus, for purely illustrative purposes, Logue's (1992) definition of sectarianism echoes many of the sentiments expressed in the section above. He advances this as:

> discrimination arising from political or religious prejudice, leading to relationships of distrust between the two major politico-religious communities. Sectarianism is not just a matter of economic, social or political consideration; nor is it simply a question of personal attitudes or behaviour. It is an historical and cultural phenomenon arising out of political and religious differences and perpetuated by group and self interest. (p.5)

The manifestations of sectarianism are all too obvious in Northern Ireland and have ranged across a combination of single identity workplaces, segregated communities, the pervasive deployment of flags and emblems as signifiers of territorial allegiance, rituals of political tradition associated with parades, paramilitary induced intimidation, civil violence and murder. The rootedness of a colonial history combined with contemporary debates around constitutional positioning and the search for reconciliation, in a society seeking with uncertainty to move beyond physical conflict, have sharpened cultural identities. The leitmotif of contestation here comprises the dilemmas of difference. Quite clearly there is a complex alchemy of religious, cultural and political attributes which on the one hand define for some people the nature of their engagement with sectarianism, and on the other hand mark out the parameters of any enthusiasm for negotiated pluralism. Experiences, influences and perceptions endure as powerful determinants of behavioural choices.

Liechty and Clegg (2001) in their analysis of the role of Christian religion in sectarianism in Northern Ireland consider three approaches for dealing with this wicked problem. The first is 'non-sectarianism' and involves working around the problem through a combination of conversational avoidance and diverting of behaviour. This is criticised as having no capacity for a positive response in times of crisis and fixing division across relationships and institutional settings through suspicious avoidance (p.25). The second approach is 'anti-sectarianism' requiring direct confrontation with its manifestations in society. However, Leichty and Clegg suggest that a process of "encounter-judge-condemn-reject-demonise-separation-antagonism" can look similar to the dynamics of sectarianism and can destroy what is good along with sectarian distortions. The third approach, which they endorse as complementary to the previous actions, is characterised by strategies of "transforming, redeeming, healing, and converting sectarian distortions" (p.26). Their vision is of a society moving beyond sectarianism which they describe as one in which:

> proactively, and without inducement or coercion, relationships of blame are giving way to taking responsibility and repentance; separating to engagement and forgiveness; overlooking to self-emptying and inclusivity; belittling, dehumanising, and demonising to mutual recognition and respect; dominating to interdependence; and attacking to peaceful co-existence. (p.340)

Within Northern Ireland the conventional sweep of public policy in working towards this vision has comprised a combination of measures related to community relations, equality and social inclusion. The next section of this chapter reviews these official discourses.

Official discourses on community relations, equality and social inclusion

Public policy in Northern Ireland has long grappled with the challenges of achieving peace and stability, growing the regional economy and targeting social need. Outwith the search for a constitutional settlement, there are three debates which have commanded central place within the policy arena over the past 35 years. These are community relations, equality and social inclusion. Each matter is inextricably linked to the other and has facilitated the creation of a raft of legislation, governance infrastructure and policy designed to encourage understanding and respect for cultural diversity, to weed-out bias, particularly within the labour market, and to reduce social exclusion.

The background to policy based efforts to improve community relations in Northern Ireland is traced by Hughes *et al.* (1998) to the outbreak of violence in 1969 and the subsequent decision by the UK Government to suspend the Northern Ireland devolved administration in 1972 because of its inability to control civil unrest. Direct rule from Westminster instigated a suite of electoral and housing reforms and, out of its commitment to examine relationships between the two dominant traditions in the region, the Government established the Community Relations Commission, along with a Ministry to oversee its work. This was modelled along similar lines to the UK Commission for Racial Equality, which dealt with race relations issues, and its membership was drawn equally from both communities. As a funding body, it supported both cross-community and single identity projects whose ultimate aim was to increase contact between Protestants and Catholics in geographically and socially segregated localities. Criticism that situating governmental responsibility within a small ministry was serving to marginalize the issue found little support at that time, but by the mid 1970s the official community relations governance had collapsed. It has been suggested that this reflected the fact that local politicians were becoming increasingly suspicious of the community development strategy being promoted by the Commission. Their concern was that a strengthened voluntary sector could provide an alternative platform for community leadership (Gallagher, 1995, p.30). Community relations made its comeback, however, in the latter half of the 1980s and was underpinned by a combination of new structures and legislation.

Firstly, a Central Community Relations Unit (CCRU) was created in 1987. This was charged with formulating, reviewing and challenging policy through the Government system and overseeing the responsibility placed on all departments to assess their policies and procedures from a community relations standpoint. Community relations considerations were linked to the delivery of key services in areas such as health, education, housing and economic development, thus extending the sweep of policy scrutiny far beyond what had been the case in the 1970s. In 1990 the Northern Ireland Community Relations Council (CRC) was established as a semi-autonomous government funded public agency to raise the profile of community relations work. Since the late 1990s CRC has extended its role to include bringing community relations work into the mainstream practices of key stakeholder organisations through a combination of cross-community and single identity projects designed to encourage cultural confidence and an appreciation of cultural diversity. The involvement of local authorities in Northern Ireland in a CCRU-sponsored Community Relations Programme initiated in 1989 was a further refinement of the assembled governance structures. By 1993, as reported by Knox *et al.* (1994) a total of 25 District Councils had joined the scheme with a budget of £1.3m. Their evaluation of the programme highlighted the enthusiasm with which councillors and officials embraced community relations as an important function of the local authority. Initiatives ranged widely across high profile community relations one-off events, single identity and cross-community development work, shared cultural traditions projects, and focused community relations such as anti-sectarian and prejudice reduction workshops.

The contribution of new legislation to community relations was made possible by the passage of the Education Reform (Northern Ireland) Order 1989. This was designed to address institutionalised segregation in education by allowing schools to opt for integrated status through a parental ballot. The legislation also provided for two cross-curricular themes to become mandatory in the teaching of most academic subjects, namely Education for Mutual Understanding (EMU) and Cultural Heritage. Essentially the aim has been to help children learn to respect themselves and others and to understand what is shared as well as what is different about their cultural traditions. Nevertheless, as observed by Knox and Quirk (2000), with less than two per cent of children attending integrated schools the low participation rate must be seen as disappointing for advocates of that approach. They additionally suggest that EMU has major questions to answer about the quality and type of experience offered to pupils (Knox and Quirk, 2000, p.84).

Notwithstanding, therefore, this considerable effort to challenge what Morrow *et al.* (2003) label as "deeply ingrained and sometimes cherished ways of behaving and believing", their authoritative analysis points to a view of community relations as seldom having gone beyond "an aim to create harmony where there was division". They provocatively conclude that:

> Community relations became a term used simultaneously to describe a vague general vision to which everyone subscribed, and a variety of haphazard practices aimed at 'harmony' that allowed everyone to remain publicly detached and knowingly cynical. (Morrow *et al.*, 2003, p.167)

The emergence of an institutionalised equality agenda can similarly be located in the top-down tradition of policy formulation and delivery within the distinctive centralised architecture of the Direct Rule state. Osborne (1996) identifies external pressures principally from Irish-American lobbyists, the Standing Advisory Commission on Human Rights and the Irish Government as forcing progress on inequality in the labour market specifically. Much of this pressure related to differential employment and recruitment rates between Protestants and Catholics and it culminated in the Fair Employment (Northern Ireland) Act 1989, which "established, what can only be seen as not only the most stringent anti-discrimination legislation in the UK, but also in Europe" (Osborne, 1996, p.185). This was followed in the same year by the establishment of a Fair Employment Commission whose duties included monitoring the religious composition of firms and organisations with more than 25 employees. A Fair Employment Tribunal was also set up to deal with cases of alleged discrimination. Additionally, in order to make workplaces less intimidating and more inclusive of cultural diversity the Government introduced the Flags and Emblems Act (1989), which outlaws the public display of political symbols in the work environment.

However, notions of discrimination and exclusion are not confined to the labour market and the Government's response to more intensive vocal external and internal political pressure led to a much wider debate about the role of the state in reinforcing fairness in related areas of public policy (Ellis, 2001). *Policy Appraisal and Fair Treatment* (PAFT) was introduced in 1994 in order to eliminate unlawful discrimination or unjustifiable inequality and to actively promote fair treatment through government policy-making and implementation. PAFT also extended the scope of equality beyond religion to include a wider range of potentially discriminatory categories such as, gender, political opinion, marital status, having or not having a dependent, ethnicity, disability, age and sexual orientation. All new public services, policies and legislation were to be formally proofed and existing services reviewed for their impact on the fair treatment of these specified categories of people.

The paramilitary ceasefires commencing in August 1994, the Good Friday Agreement of April 1998 and the subsequent establishment of a Northern Ireland Assembly and devolved government brought PAFT out of its previous policy domain and re-framed it as a legally binding regulatory practice. The Northern Ireland Act 1998, which provided the legal interpretation of the Agreement, set down formidable equality duties on all public bodies in the region. This gave legislative weight to the concept of policy proofing and *ex ante* appraisal, and gave legal force to the recognition of nine equality categories under Section 75 of the Act. Under its provisions due regard must be given to the need to promote equality of opportunity:

(a)   between persons of different religious belief, political opinion, racial group, age, marital status or sexual orientation;
(b)   between men and women generally;
(c)   between persons with a disability and persons without;
(d)   between persons with dependents and persons without.

In carrying out their functions, public authorities must also "have regard to the desirability of promoting good relations between persons of different religious belief, political opinion or racial group" (Section 75 (2)).

Government Departments have been obliged to prepare Equality Schemes that show how these objectives would be met through current policies and Equality Impact Assessments (EQIAs) have been introduced to proof all new programmes against the needs of the nine groups identified in Section 75(1). Equality Schemes describe the specific functions and policies of the relevant public body, the policies to be subjected to EQIAs, and a timetable for implementation. The Schemes devote considerable attention to monitoring, publicity, consultation and complaints procedures. However, training and skills development has been aimed at achieving a greater understanding of the operation of the Scheme rather than a wider understanding of conflict and the role of policy in reducing inequalities *per se*. Like the Equality Schemes themselves, EQIAs have, in some cases, been characterised by a reductionist model designed to comply with the legal minimum rather than developing a proactive stance on the causal relationships explaining inequalities and how they are manifest within specific policy domains.

As with PAFT, the unfolding peace process and the 1998 Good Friday Agreement also provided the impetus for Targeting Social Need (TSN) to be evaluated and reworked as the cornerstone of the Government's commitment to social inclusion. In 1991 the then Secretary of State for Northern Ireland, Peter Brooke, had announced that TSN was to become a public spending priority in the region. The implications of this were that resources would be skewed towards those areas of greatest social need in order to improve the position of the most marginalized people in Northern Ireland. It was anticipated that TSN-led decisions would reduce unfair social and economic differentials, and promote equality for all sections of society. Research by Quirk and McLaughlin (1996) was subsequently scathing on the TSN performance of central government departments. Their analysis indicates that while a number of local development initiatives gave profile to TSN, the structural pattern of public expenditure remained largely the same The remodelled initiative of July 1998, called *New Targeting Social Need* (NTSN), thus represented a timely response to that critique. This had a number of key elements: tackling unemployment, addressing social need (in health, education and housing), and coordinating the actions of Government Departments through a new commitment to Promoting Social Inclusion (PSI) (NTSN Unit, 1999). Departmental Action Plans have again translated these themes into three-year programmes and, in particular, identified poverty reduction targets for specific policies and initiatives. But these vary radically in their level of detail, style and the way in which NSTN commitments are to be tested and evaluated. Some Action Plans make vague commitments to researching the causes and effects of deprivation. The monitoring of conditions over time is a recurring theme in most Departmental approaches. Some make specific commitments on inputs, such as where activities will be located and resources will be invested, whilst others try to explain how the outputs of policies will result in some form of social and economic closure.

Accordingly, official discourses in relation to community relations, equality and social inclusion have relied heavily on structures, legislation, procedure, rational information systems, funded programmes and audits. The problem here is that organisational perspectives on community relations, equality and social inclusion really do need to be deeply embedded in the structures, processes, values and routines of the relevant agencies in more meaningful ways if they are to be effective. EDI, we argue, presents a framework for this process of systemic adjustment by nurturing the capacity to think creatively about how to move these agendas beyond their narrower bureaucratic boundaries. We believe that EDI provides the potential to secure what Liechty and Clegg (2001) call "transformed patterns of relationship" in situations where there are embedded cultures of "blaming, separating, overlooking, belittling, dehumanising, demonising, dominating and attacking" (p.340).

Equity – Diversity – Interdependence

The principles of equity, diversity and interdependence (EDI) in Northern Ireland derive, in the first instance, from research and policy applications within the field of community relations. The Community Relations Council Strategic Plan 1998-2001 defined these principles as follows:

- equity is understood as a commitment at all levels within society to ensuring equality of access to resources, structures and decision making processes and to the adoption of actions to secure and maintain these objectives. Equity is about equal opportunities;
- diversity can be seen in the ever changing variety of community and individual experiences. Respect for diversity affirms the value which can be derived from the existence, recognition, understanding and tolerance of difference, whether expressed through religious, ethnic, political or gender background. Diversity is about being free to shape and articulate identities;
- interdependence requires a recognition by different interest or identity groupings of their obligations and commitments to others and of the inter-

connectedness of individual/community experiences and ambitions leading to the development of a society which is at once cohesive and diverse. Interdependence is about quality of relationships.

The Community Relations Council (1998) takes the view that "civil society depends on a shared discourse which recognises and affirms differences, but allows these to exist in constructive relationships with each other. For this to happen, initiatives at all levels must be able to integrate these principles into their work in appropriate ways". (p.6)

The contribution of Eyben *et al.* (1997) has been especially influential in articulating the value of these principles for societal relations in Northern Ireland. They point out, however, that while activities have long focused on inter-personal and inter-group encounters, the process efforts have been tied more to the community level rather than also embracing government body and institutional stakeholders. A legislative and policy superstructure is now in place to deal with aspects of equity, diversity and interdependence, but the extent to which this is being operationalised in a truly transformative manner at the middle, delivery-oriented level, comprising a multiplicity of Northern Ireland wide and area-based service organisations, remains unclear.

One initiative, which is designed to give support to organisations in dealing with the challenges of mainstreaming the principles of fairness, valuing difference and mutual respect, is being led by Counteract and the Future Ways Programme at University of Ulster. This is called *The EDI Framework* and comprises an approach for the animation of change with the following characteristics:

- a process to assist the leadership and staff within an organisation to deepen their understanding, confidence and skills in managing relationships based on difference and equity;
- an internal self assessment framework which allows an organisation to plan, implement and measure against previous internal benchmarks and, if the information is available, in comparison with other similar organisations;
- a foundation process and a set of values underpinning and enhancing existing standards and quality management initiatives. (Counteract and Future Ways Programme, 2001)

Our research, as reported in this book, is complementary to that work. It seeks to examine "provider - consumer" interactions in terms of values, relationships and work practices from an organisational perspective, but to do so in a manner which gently moves the official discourses of equity, diversity and interdependence well beyond anti-sectarianism and reconciliation, towards a more complete engagement with social inclusion. Morrow *et al* (2003) describe this challenge in the following fashion:

> Rather than coercive legislation, which requires conformity with pre-ordained legislative outcomes, the requirement is for measures which support the development

of a culture of learning and development which encourages innovation and commitment in pursuit of an agreed vision and values. The key measure of success in such policy is in the growth of new capacity to deal with difficult but real problems rather than the absence of surface difficulties which leaves underlying issues untouched. (p.180)

The application of EDI has, therefore, wide relevance for the discussion of social and economic wellbeing and reaches across the mosaic of public, private, community and voluntary bodies that comprise contemporary governance. The primary research reported in Section Two, involving a range of service organisations impacting on rural society in Northern Ireland and using several perception lenses (for example, political affiliation, gender, religion, race, sexual orientation, disability, age, employment status) fits well with this re-conceptualisation.

This analysis, in turn, may require some re-working of the interpretation of equity, diversity and interdependence. Thus, for example, in relation to the equity principle it may be necessary to overtly build in "inter-generational equity" which implies a long term approach to local capacity and asset building, rather than the political expedience of short term goals and targets. In relation to diversity it may be appropriate to overtly recognise diversity of method within organisations rather than just individual and community diversity. And finally, in relation to interdependence there may be a need to give greater weight to local engagement, from design to delivery, while being mindful of necessary linkages with local democracy. These considerations would seem critical in forging effective responses to the driving processes of social exclusion discussed at the outset of this chapter and ultimately move the relevance of equity, diversity and interdependence into the sphere of social capital formation.

Conclusion

This chapter has served to define the conceptual basis for our research as dialogues between governance and people which are constructed around ideas of equity, diversity and interdependence. The departure points comprise social exclusion, social prejudice and sectarianism. These have been responded to by a complex suite of policy interventions which have been constructed within the state discourses of community relations, equality and social inclusion. These official discourses are each identified with inevitable implementation gaps that underline the need to move beyond political, legislative and bureaucratic responses for the achievement of truly systemic relational changes. This is not to argue that these responses are irrelevant. The important point here is that the embedding of an EDI framework more completely into all the routines of governance will deepen the scope for societal transformation. In Part Two of this book we move on to present our research findings. We commence with a brief outline of the changing governance context in rural Northern Ireland in order to set the scene for that investigation.

PART TWO:
BASELINING EQUITY, DIVERSITY AND INTERDEPENDENCE IN RURAL NORTHERN IRELAND

The Governance Arena of Rural Northern Ireland

Introduction

While Northern Ireland is synonymous with particular societal divisions, its experience of a longstanding rural – urban dialectic is shared in common with other parts of Europe and North America. In part, this is a problem born out of definitional uncertainty which categorises the rural as either open countryside, or a combination of nucleated and dispersed settlement at the sub regional scale. In each case, however, the spatial baseline remains the pattern of major urban locales and as a classification (Cloke, 1978) of rural areas illustrates, using distance as a key variable, the hegemony of the urban is difficult to undo. The identification of pressured rural areas, accessible rural areas and remote rural areas mirrors the conventional codification of core and periphery in regional planning practice and takes the urban as the essential departure point for analysis.

Arguments of equity, capacity and empowerment as expressed through community led local development and area based partnership governance have sought, however, to challenge the marginalising of rural society. One effect has been that government policy makers have had to adjust their perceived wisdom on differential patterns of locational advantage and adopt a more interactive mode of planning and investment allocation. But harnessing the capacity of rural people to engage effectively and on a sustained basis with public officials requires the adoption of collaborative strategy building and initiative-based development processes. This chapter deals with these matters. It commences by describing the scale and character of rurality and then revisits in more depth the current debates on social deprivation, equality and societal division as they relate to rural Northern Ireland. The discussion moves on to outline the emergence of rural development as a prominent feature of public policy in Northern Ireland and concludes by describing the complex governance arrangements which have been in place to secure a wide range of development outcomes. It is within this context that the detailed analysis in the remainder of this section of our book is situated.

Rural scale

From the perspective of town and country planning, rural Northern Ireland has been defined by the Department of the Environment (1993) as everywhere within the region outside the Belfast metropolitan area and Derry/Londonderry (Figure 3.1). In

1991 some 60% of the region's population (950,000) lived in the area defined as rural Northern Ireland, of which some 350,000, or 22%, lived in the open countryside. This expansive definition of the rural comprises a combination of regional towns, small towns, villages, cross-roads settlements and dwellings in open countryside and hints at an interpretation which is mindful of a strong functional interaction across settlement components. More recently, the Department of the Environment (1998) has sought to sharpen that focus for regional planning purposes through its advocacy of the perspective of a family of settlements which distinguishes between the regional towns and what is labelled as "the Rural Community". For 1996 this redefined rural population was estimated at 652,250 amounting to 39% of the population of Northern Ireland. The important point here is that, by either definition, the rural constituency is large, an observation which is underlined by the fact that the Belfast metropolitan area comprised 36% of the Northern Ireland population, with Derry/Londonderry, Craigavon and the other Regional Towns making up the balance of 25%. The rural constituency, in short, contains a diversity of communities within which people live and work.

Physical planning policy at its most basic is concerned with helping to shape the future distribution of population across a territory by managing development pressures and processes. That policy may change over time, in line with regional strategies which seek to promote containment or expansion in urban areas and strict control or more relaxed stances on development in rural areas. Within Northern Ireland the Rural Community comprising a dispersed settlement pattern of small towns, villages, cross-roads developments and dwellings in open countryside has demonstrated an accelerating growth rate over 3 time periods since 1971. As illustrated in Table 3.1, significant growth has taken place in the small town, village and small rural settlement components of the Rural Community; over the period 1971 to 1996 the increase was 45% whereas, in contrast, the equivalent for the open countryside was 3.4%. This runs counter to a populist perception of the open countryside being swamped by housing development.

Table 3.1　Population Trends in the Northern Ireland Rural Community

The Rural Community Component	1971	1981	1991	1996
Small Towns (under 10,000 pop)(1)	134,482	159,071	187,874	204,100
Villages and small settlements	68,417	76,034	83,646	89,300
Open countryside	346,954	344,274	349,667	358,850
Total	549,853	579,379	621,187	652,250

(1) Excludes the towns of Ballycastle, Ballymoney and Magherafelt which are Regional Towns

Source: Department of the Environment (1998) Shaping Our Future - The Family of Settlements Report, p9.

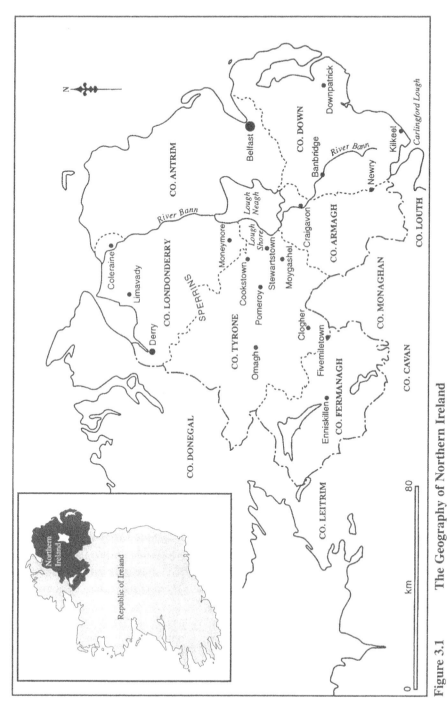

Figure 3.1   The Geography of Northern Ireland

However, the intensity of development activity implicit in this analysis of population change within the Rural Community does vary across Northern Ireland. Within the Belfast City Region, data from the Department of the Environment (1998) highlight population growth for the Rural Community of some 32% between 1971 and 1996 compared with 10% in the rest of Northern Ireland over the same period. Whereas the population of the Belfast metropolitan area declined from 653,540 in 1971 to 591,950 in 1996, the small towns and villages within the Belfast City Region increased from 91,779 to 143,700 (+51,921) over the same period, with considerable impact on the scale and character of some of these settlements. Equally interesting is the recording of population growth of 13% in the open countryside of the Belfast City Region which increased from 118,760 in 1971 to 133,750 in 1996. This growth is all the more striking given the longstanding implementation of development restraint policies throughout much of this countryside. Within the remainder of rural Northern Ireland the population of the small towns and villages increased from 111,120 to 149,700 over the period 1971-1996, but the open countryside component actually fell by 2% from 228,194 to 225,100 persons. In short, the trend-based evidence points to varying patterns of place based population growth and stability across rural Northern Ireland.

Rural imagery

The rural in Northern Ireland, as elsewhere, is very much a social construction whose meaning varies across individuals and across interest groups. For some it is the simplistic essence of everything that is good in society – neighbourliness, fresh air, or great scenery. It is thus the antithesis of everything that is perceived as less wholesome (and urban) – high density housing and anonymity, traffic congestion, poverty and violence. As a social construct the rural comprises a complex melange of place, culture, and identity and which, as suggested by Duffy (1997), may for most people be more myth than reality. Such representations, however, do serve to set it apart. The accompanying imagery, filtered from historical and contemporary realities, can invoke a romanticism and nostalgia, which at best is but a partial illustration of circumstance. Photographs can be an especially powerful influence on perception and behaviour as frequently portrayed by the branding of Northern Ireland for tourism promotion purposes as a place of bucolic harmony. But behind this gentle illusion of a normal society, alternative and unsaid interpretations of rural living are equally feasible: the isolation of the elderly, the marginal viability of the small farm, the depletion of community services. There are real issues of equity in rural Northern Ireland.

The cultural landscape, therefore, is a potent reservoir for informing analysis and policy prescription with a similar iconography capable of being used for different public agendas. The case of rural housing in Northern Ireland is a useful example of this multiple meaning. In a richly illustrated investigation into the constructional and typological aspects of rural houses, Gailey (1984) makes a genuine plea for the conservation and restoration of vernacular dwellings for their own sake. But a subsequent and again richly illustrated publication from the Northern Ireland Housing Executive (1990), as part of its rural housing policy review, questions the picturesque postcard view of the thatched cottage which masks a lack of amenities and isolation. Its research identified a litany of rural housing problems at that time, for example, approximately one dwelling in 12 lacking basic amenities, more than 3,000 dwellings without mains electricity and a similar number not connected to a mains water supply, and some 68% of all vacant rural dwellings regarded as unfit. In 1991, a policy statement *The Way Ahead* (Northern Ireland Housing Executive, 1991) argued not only for refurbishment aid, but also replacement grants. This differential commodification of rural housing into built heritage and unfit shelter is a depiction of broader debates about the nature of rurality in Northern Ireland which, only in recent years, have focussed on the reality of social deprivation. A deeply divided society provides an overlay of contested territory and relationships.

Rural deprivation and targeting social need

As an alternative discourse in the interpretation of rural Northern Ireland, social deprivation is not only more recent, but also represents a cultural iconography representative of more fundamental perceived power imbalances between Belfast with its hinterland and the rest of the region. This has been poignantly embodied in the slogan 'West of the Bann' (Figure 3.1) which signifies separation by a river of a disadvantaged rural sub-region in the west from its more prosperous and urban counterpart in the east. Unravelling the spatial configuration and complexity of this disadvantage has moved on from days when the rural problem was solely aligned to the hardship of the small farmer, whose interests were represented by a union with credentials for entry into the policy community of government. A conventional analysis ran as follows:

> Finally, it should be borne in mind that while commercial farming and state regulation of agriculture have been the dominant trends in the last few decades, skills and attitudes that are far older have been by no means lost or abandoned. The small farmer inherits an indefinable quality of stockmanship which is perhaps his greatest asset, and even if he can no longer do what he likes he generally likes what he does. He is not given to spending money which is hard-won and it tends to be valued far more than the commodities that it can buy. There are many parts of the country where modesty and other peasant attributes are regarded as the highest virtues and ostentation is deplored. In the face of these attitudes capital investment and improvements are not readily undertaken...The problem of improving the quality of rural living and providing modern amenities is further increased by the extreme dispersion of the farmsteads. (Symons, 1963: p.22)

The identification of rural diversity and the persistence of a broader rural deprivation owes much to the seminal research of Armstrong et al (1980). Six categories of rurality were noted ranging from the periphery of the Belfast Urban

Area (Category 1) to wards which displayed a marked dependency on primary sector employment, contained the highest levels of over-crowding and unemployment, and had the lowest provision of household amenities (Category 6). The latter are defined in the report as "rural problem areas", accounting for two-thirds of the land area of Northern Ireland and a 1971 population of 230,000. A subsequent reworking of this analysis by Caldwell and Greer (1983) which incorporated those Category 5 wards adjacent to Category 6 offered a more complete portrayal of 'peripheral rural areas', the principal effect being to form a monolithic block which stretches from North Antrim, through to almost the entire west of the region, the Armagh/Monaghan borderlands, South Armagh and South Down (Figure 3.2). This equates with a population of 276,830 in 1981, almost 19% of the population of Northern Ireland at that time. Apart from the already noted correlation of rural peripherality with scenic amenity, it is notable that these areas (with the exception of North Antrim) either directly abut, or run close to the Border between Northern Ireland and the Republic of Ireland. Its proximity in some cases creates peripherality, as in the north-west, or at the very least adds to it, with every peripheral rural area in Northern Ireland having its counterpart in Counties Donegal, Leitrim, Cavan, Monaghan and Louth (see Haase et al, 1996: p13).

Figure 3.2   Peripheral Rural Areas in Northern Ireland

While these analyses have been updated and refined during the interim by Robson *et al.* (1994) using *inter alia* 1991 Census data, and most recently by Noble *et al.* (2001), the recurrent pattern of deprivation remains the peripheral rural areas of the west, south and north east of the region, together with a suite of urban wards in Belfast and Derry/Londonderry. At an aggregate scale, analysis of the 1991 census also suggests an ethnic geography of Northern Ireland which has become more sharply demarcated with Catholics forming a substantial majority in all of Counties Fermanagh, Tyrone, and Londonderry, north-east Antrim and parts of Down (Graham, 1997 pp.201-202). The success of Sinn Fein and the Social Democratic and Labour Party in winning parliamentary seats within this territory in the 2001 General Election underlines the depth of the political transformation which has occurred across much of rural Northern Ireland.

The public discourse on rural disadvantage as conveyed by its spatial representation has become a powerful metaphor of key socio-economic differentials which have built up over time between the competing traditions of Northern Ireland. As discussed in Chapter 2 'Targeting Social Need' (TSN) became one tool aimed at changing the policy making process across the profile of public expenditure in the pursuit of greater equality of opportunity. The symbolism of TSN may, however, have outweighed its impact since, as already noted (Quirk and McLaughlin, 1996), there is little evidence that it had a substantial influence on the decision-making and spending of government departments. Specifically in the case of the then Department of Agriculture (DANI), the lead department for agriculture and rural development in Northern Ireland, mainstream programmes were identified as having been untouched by TSN, with only rural development, 'a small new area of activity', in any way related to this priority. Given, however, that this accounted for 6.5 per cent of DANI's expenditure and depended heavily (75 per cent) on European Union funding, it was concluded that this "programme is clearly marginal and separate from the department's overall activities" (Quirk and McLaughlin, 1996: p.166).

It is significant, therefore, that Government announced the development of New TSN in 1998 as a further commitment of its intention to respond more effectively to the needs of the most disadvantaged people in Northern Ireland. New TSN is not an initiative with its own budget, but is seen as being a cross cutting and targeting feature of existing spending programmes, not least related to the priorities set out in the annual Programme for Government. A commitment to publish New TSN Action Plans and Annual Reports on progress could arguably be regarded as an attempt to demonstrate performance and transparency in the face of that earlier criticism. At the same time there is scope for a more cautious interpretation of outcomes as observed by Ellis (2001). His assessment of the action plans of two government departments with a direct interest in land use planning suggests that only minor tinkering has taken place with existing policies, rather than the more fundamental realignment of objectives and expenditure required by the New TSN rhetoric:

> Any analysis of these documents can only conclude that they consist mainly of a reiteration of the conventional wisdom of modernist planning discourse clad in platitudes of achieving social inclusion. (p.403)

Dealing with diversity

The passage of the Northern Ireland Act (1998) marked a significant effort by Government to introduce provisions designed to secure deeper appreciation for diversity and to link this with the broader goal of inclusion. Section 75 of that legislation, to restate its content, requires that listed public authorities in carrying out their functions have due regard to the need to promote equality of opportunity between: persons of different religious belief, political opinion, racial group, age, marital status or sexual orientation; men and women generally; persons with a disability and persons without; persons with dependents and persons without. Public authorities are also obliged to have regard to the desirability of promoting good relations between persons of different religious belief, political opinion or racial group. The implications for rural Northern Ireland are that accepted spatial preferences for intervention based on deprivation are now complemented by the insistence that policy impacts are screened for equality of opportunity against the realities of difference in rural society. In carrying out this obligation the Equality Commission (1999) has suggested that it is necessary, for example, for consideration to be given to any evidence of higher or lower participation or uptake by different groups, evidence that different groups have varying needs or experiences of a particular policy, or whether there is an opportunity to promote better community relations by altering the policy, working with others in Government or in the wider community.

Thus research commissioned by Rural Community Network (Shortall and Kelly, 2000) on gender proofing the reforms of the Common Agricultural Policy (CAP) is located within this legislative context for diversity and provides a significant illustration of differential policy impacts on women and men. Key issues include: the underestimation of the work that women do on farms, an early retirement scheme for the two adults retiring rather than just the male "farmer", training provisions for women due to changes brought about by the CAP reforms, stress levels linked to reduced incomes together with a combination of childcare and farm work responsibilities, and the under-representation of women in rural development. The central finding of this seminal research is that the broad social consequences of CAP reforms are not being sufficiently considered by policy makers.

Again, as introduced in the previous chapter, a particular manifestation of the absence of respect for diversity in Northern Ireland is social prejudice and which all too often finds frequent expression as sectarianism. Rural Northern Ireland has suffered its share of sectarian tension and violence, thus creating a patchwork rural geography of inter and intra settlement apartheid. This divided society has been characterised by a community infrastructure constructed around mutually exclusive single identities, villages and small towns cut off from their natural hinterlands by cross border road closures, housing estates with emblem reminders of community territoriality, and central areas marked by property destruction and an aversion to private investment. Research by Murtagh (1996, 1998) demonstrates that the social processes which have created 14 'peacelines' in Belfast do not stop at the greenbelt and that community attitudes in rural Northern Ireland need to be understood in the context of the political and ethno-religious group experiences of living through violence as well as perceptions of ethnic sustainability.

In short, the challenges facing many small towns, villages and dispersed rural communities in rural Northern Ireland are both complex and multidimensional. There are contested policy agendas and priorities, contested spaces and contested traditions which point to the existence of different rurals for different people. Moreover, rural society is itself highly differentiated across groups and the groupings themselves need also to be disaggregated on the basis of diversity and inclusion. In other words, there are multiple and overlapping identities and conditions which impact on individuals. Thus, for example, the experience of women in rural life can be stratified on the basis of farming or non farming background, age, race, sexual orientation, family/marital status, disability, religion, political opinion and so on. The inequalities, which express themselves as varying forms of prejudice may be related, for example, to isolation, non-recognition, poverty, and powerlessness. Social prejudice, in general, and sectarianism, in particular, are commonly acknowledged as being difficult issues to talk about because of sensitivity and as being difficult issues to quantify because of silence. The more local and the more personal conversations become, then the more problematic it is to confront the realities of diversity and exclusion. Policies brought forward within the arena of rural development governance have belatedly sought to make a contribution to the wider quest of restoring community confidence and building stability.

Community led local development in rural Northern Ireland

Northern Ireland is marked by a large and vibrant community/voluntary sector. Birrell and Murie (1980) have pointed to some 500 community groups and associations in existence in 1975, while more recently the Northern Ireland Council for Voluntary Action (2002) has estimated the combined total of voluntary and community organizations at 4,500. Their interests are wide ranging and embrace planning, advocacy, service delivery and job creation in a manner which is complementary to the work of public bodies and the private sector. The sector employs some 29,000 people and more than 72,000 volunteers give their time freely. Community development processes have been central to these tasks and in the difficult context of Northern Ireland these have sought additionally to construct necessary bridges across a deeply divided society.

As reported by Lovett et al (1994) the community development movement emerged during the 1970s out of a more radical action based ethos which proclaimed that people should have a greater say in the decisions affecting their everyday lives. In Belfast and Derry, for example, successful campaigns were mounted against housing redevelopment and road proposals. By the 1980s many community groups had been absorbed as part of the broader state welfare system, with earlier oppositional stances to public policy tempered by a co-optive engagement based on responsibility and funding. While the initial locus of activity was essentially urban, this has been dramatically counterbalanced during the past decade by an increasingly organised rural constituency which has suffered equally from the worst effects of deprivation and violence. Herein, perhaps, is one of the paradoxes of Northern Ireland: the coexistence of a blossoming civil society in the mire of deep societal division. A defining characteristic of this participative democracy is that it involves people in decision-making thus nurturing a social ability to collaborate for shared interests. Robert Putnam in his seminal 1993 book *Making Democracy Work* argues that without these norms of local reciprocity and networks of engagement the outlook in any society is bleak. Outcomes include clientelism, lawlessness, economic stagnation, and ineffective government. Strong participatory citizenship whereby people are involved in planning and in implementation, in facilitative leadership roles and in creating better futures for their own communities is a necessary condition for avoidance of those ills. A benign institutional environment is, however, an overarching requirement. The recent experience of community led local development and local partnership governance is illustrative of this linkage between rural development and civil society.

Increased interest by the European Commission and lobbying at a local level prompted by the recommendations from a pilot rural community action project led the then Secretary of State for Northern Ireland, Peter Brooke, to appoint an Inter Departmental Committee on Rural Development (IDCRD) in 1989. Its brief was to advise him on the best way of carrying forward action to tackle the social and economic problems in the most deprived rural areas in Northern Ireland. The committee reported in December 1990 with a panel of recommendations. In summary these were: (1) DANI should become the lead department with responsibility for promoting the development of the most deprived rural areas; (2) a permanent inter agency forum should continue in operation to oversee progress; (3)an independent Rural Development Council (RDC) should be established, sponsored and funded, at least in its early years, by Government. Its staff would be charged with responsibility for community led rural development and the organisation would also provide a forum for discussion between communities; (4) DANI should appoint a small team of local coordinators from across the public sector to bring together the responses from all public sector agencies to plans and projects located in the most deprived rural areas; and (5) a specific but relatively small fund for rural development should be set up to supplement mainline programme expenditure.

These recommendations were accepted in principle by government and formed the basis of its Rural Development Initiative launched in 1991. Thus from the outset, a new agency, outside government but dependent upon it for funding, assumed a pivotal role in facilitating community based social and economic regeneration. The RDC determined that its entry to this field had to be through generic community development in the first instance and in late 1991 commenced the process of appointing a team of 6 community officers. The 1992-1995 Strategic Plan of the RDC defined their role as comprising: stimulating awareness of the value of collective action at local levels; building confidence in ability to bring about improvement; encouraging marginalised groups to participate; and promoting the adoption of good working practices. This community development support was accompanied by financial assistance to facilitate group formation, to participate in training and education activities and to purchase technical services, for example, in regard to development strategies and project business plans. The RDC also articulated in its first Strategic Plan a commitment to providing a research and information service, and a willingness for cooperation and consultation with a wide range of rural interests. However, there is no mention in the documentation of the role identified by the IDCRD for RDC to provide a discussion forum between communities. This task was addressed by the establishment in 1991 of Rural Community Network (RCN), a voluntary organisation which emerged out of the earlier Rural Action Project with a mission to identify and voice issues of concern to rural communities in relation to poverty, disadvantage and community development.

The achievements of the Rural Development Programme over the decade from 1991 have been identified by the Department of Agriculture and Rural Development (2001) as comprising the creation of over 1,000 jobs, the maintenance of over 900 existing jobs, support to 3,400 individual projects, 450 new rural business start-ups, 2,000 people completing training and further education courses, and 450 community groups involved in rural development. Notwithstanding, the difficulties attached to the multiple counting of programme outputs on the basis of cocktail funding, the evidence is that an energised rural constituency has responded to the opportunities held before it. However, an annual expenditure of some £8 billion by the government in Northern Ireland sets the modest commitment of the rural development programme, as noted above, in context, albeit that other budget items do benefit rural society.

Partnership governance and rural development in Northern Ireland

Partnerships are now embraced in advanced capitalist societies as a service delivery mechanism. While not being a panacea for solving local development problems, partnerships can be effective mechanisms to improve relationships among multiple stakeholders and to bring together human and financial resources from a variety of funders to achieve common objectives. The concept implies a change in the nature of governance, not as an alternative to elected representation, but with the state taking a less pronounced role in dealing with complex problems such as urban and rural disadvantage or social exclusion. Moreover, the notion of partnerships converges easily with political pressures for a reduction in state activity and increased responsibility at the local level.

It is not surprising, therefore, that the corporate strategies, operational plans and annual reports of government departments and agencies within Northern Ireland should resound with approaches to creative collaboration. Indeed the rural development arena abounds with local partnerships. Elected representatives, public officials, the business and trade union sector, and community and voluntary interests have collectively engaged in a more grass roots approach to resolving local problems. The sheer scale of partnership governance for rural development is highlighted by the following structures which were operating during the 1990s:

- Under DANI's Rural Development Programme, in place through to 1999, there were 8 Area Based Strategy Action Groups, each of which was resourced with a budget of £1million in order to lever in additional funding for regeneration projects within disadvantaged rural areas. The first Area Based Strategy was launched in South Armagh in December 1995 within a border region which has suffered a negative image due to terrorism. The strategy covered small business development, agriculture, environmental management and tourism development and was implemented in association with a locally based cross-community regeneration group.

- There were 15 LEADER 2 local action groups in which District Councils played a prominent leadership role and whose brief comprised the giving of support to innovative, demonstrative and transferable rural development initiatives. LEADER+ has provided a measure of continuity for local action groups through to 2006.

- District Councils were able to establish local economic development partnerships which were concerned with the delivery of economic development measures contained within the EU Structural Funds Northern Ireland Single Programme for the period 1994-1999.

- Within each local authority area there was also a District Partnership established under the EU Special Support Programme for Peace and Reconciliation and with a funding remit which comprised social inclusion, rural and urban development, productive investment and employment. In 2001 the 26 District Partnerships were rebranded as Local Strategy Partnerships.

- The Department of the Environment in partnership with the International Fund for Ireland (IFI), District Councils and rural communities contributed to small town and village renewal through a portfolio of Community Regeneration and Improvement Special Programme (CRISP) projects. These comprise the blending of a core community business scheme, environmental improvements, and grant aid for private sector development.

Common to all these illustrations of partnership governance-in-action is (1) the critical involvement of local authorities in lead or support roles; (2) the preparation and implementation of locally prepared strategic plans; (3) a dependency upon public funding from EU, IFI and mainstream sources; (4) a high level of voluntary participation; (5) an appreciation that the rural development challenge is multi-dimensional extending across economic, social, environmental and infrastructure needs; and (6) a policy preference, thus far, for a project driven implementation approach in which community groups are a central delivery mechanism. There is, however, growing acceptance that this highly differentiated rural development arena has been rapidly 'crowding out' through the presence of multiple delivery agents which extend beyond the partnerships identified above, to include other organisations such as local enterprise agencies, the Rural Development Council and Rural Community Network (Northern Ireland Economic Council, 2000; Hart and Murray, 2000; PriceWaterhouseCoopers, 1999). Moreover, with the benefit of hindsight, serious questions have been posed about the wisdom of a policy imperative which almost exclusively has advocated large scale capital support for community led projects. Difficulties related to project management, profitability and the repayment of loan capital have surfaced in a number of instances and have prompted challenging scrutiny by the Northern Ireland Audit Office (2000) and the Public Accounts Committee of the Northern Ireland Assembly (2000). The bottom line questions to be debated after almost 15 years of activity are the sustainability of this local scale participatory planning and development effort and its future effectiveness, if geared at this high level and when set against the prospect of a normal Northern Ireland, which increasingly will be driven by private sector investment, and a reducing EU grant aid environment during the decade ahead. The latter scenario has prompted the establishment in 2003 of a Task Force by government to review the long term funding of the community and voluntary sectors.

In the latter part of the 1990s Government released a suite of prominent statements on regional physical and economic planning in Northern Ireland, which also included preliminary proposals for housing growth, transportation infrastructure and hospital services. In late 2001 a new Rural Development Programme to cover the period to 2006 was launched (DARD, 2001) and the Northern Ireland Assembly finally approved a new regional development strategy for Northern Ireland (DRD, 2001). Their preparation and implementation is inextricably linked to a much more active, better organised and astute rural constituency. This engagement spans the local and regional scales and while community and voluntary led initiatives can assist with the reshaping of rural space at a micro level, rural constituencies acting collectively can also influence broader patterns of change.

Conclusion

The analysis in this chapter has sought, firstly, to unravel the multiple threads of policy meaning which are wrapped around the concept of *the rural* in Northern Ireland. What is clear is that rurality has become a powerful metaphor for claims related to spatial equity, bottom-up development processes, and intergenerational sustainability. An outside perception of rural areas as solely an environmental resource is being confronted by the internal reality of diverse and active community interests for whom place, culture and identity are increasingly powerful signifiers.

Chapter 4

Perceptions of Diversity and Inclusion Among Northern Ireland-wide Service Organisations

Introduction

The data in this chapter deal with perceptions of rural diversity and inclusion among a number of Northern Ireland-wide service organisations. The research is presented under the following main headings: awareness and experience of rural exclusion, division and integration; understanding of and commitment to equity, diversity and interdependence; Equality and New Targeting Social Need imperatives and the EDI implications; training in prejudice awareness and reduction; dialogue with and among rural people; and collaboration with other organisations including Rural Community Network. Background notes on the organisations identified in this and following chapters are presented in the Glossary of this book.

Awareness and experience of rural exclusion, division and integration

A common theme pursued by a number of interviewees was the stated perception that community led rural development has long been regarded by many Protestant groups as a Government led initiative to give exclusive support to Catholic/Nationalist areas. This has additionally spilled over into how some of the membership and service organisations promoting rural development such as the Northern Ireland Agricultural Producers Association (NIAPA) and the Ulster Farmers Union (UFU) have been perceived. The following comments illustrate this situation:

> NIAPA is perceived as a bunch of Nationalists and the Ulster Farmers Union is perceived not just as a bunch of Unionists, but also Presbyterians. Of course this is nonsense. But if that is the perception, then somehow NIAPA and the UFU exude these feelings.

In a similar vein RCN, for example, was described by one interviewee as being perceived at its establishment as "Catholic, west of the (River) Bann oriented and male dominated". This may well reflect the initial thrust of the Northern Ireland Rural Development Programme which was spatially concentrated but has since moved on to include wider rural geographies. RCN on the other hand has sought to counter this perception through serious internal discussion and promotion of its role in challenging poverty, promoting inclusion and tackling sectarianism throughout rural Northern Ireland. It regards itself, for example at Board level, as having successfully moved up a performance curve from its establishment to embrace a gender and religion balance. However, an important related observation with wider applicability was expressed in the following way:

> You can never take your eye of the ball; you may think you have achieved something, but actually because you just watch the numbers, you may not have secured a broader shift in the power base. Inequality in gender, age and religion can still manifest itself at any one time in the process of elections to the Board. That does not mean of course that at the next election this pattern will repeat itself. We have started to be proactive, for example in regard to gender. We have been considering the type of training that might be required by way of corrective action so that next time there will be a better profile.

In short, notions of equity can change over time and comprise a process of constant re-negotiation which is shaped by wider changes in social values.

Interviewees reported a number of different ways by which social exclusion manifests itself. This was identified as having both a spatial dimension, for example, being able to avail of public services in remote rural areas, and a prejudice dimension associated with a range of societal groups. One of the most graphic illustrations of prejudice relates to the *Travelling community* and who, within the context of Northern Ireland as a divided society, was described as follows:

> It is the only thing that unifies people in Northern Ireland – the hatred of Travellers. This spans religion, politics, social class, whatever. They are not seen as Irish Travellers with their own culture. They are seen as a problem, because they are parked on the side of the road, or they are parked in areas for a long time with no facilities. The State reflects this attitude that there is no welcome for them.

This observation relating to the institutional separation of Travellers' welfare was outlined in the following manner:

> The Housing Executive has responsibility for group housing and after 2002 for serviced sites. But nobody has responsibility for transit sites or short stay sites. And yet Travellers are included in the Race Relations Act as a nomadic culture. The attitude seems to be that if we give them good enough services then they will settle down and they will not want to be Travelling any longer. The conventional equality deal is based on us all being settled; it is set around our culture rather than Travellers' culture. Because of this there is institutional discrimination towards Travellers. This is not deliberate but there is the feeling that Travellers should change; their lifestyle is seen as wrong, as deviant, not the same as ours, lesser than ours. Thus if Travellers move into a house, if they take up public services – that's regarded as equality from the agencies' perspective but it involves Travellers changing their culture. There are Travellers who have moved into houses, who live in extended family groupings and go off for five months of the year and then come back. But the education system is

not based that way. In regard to the health system, Travellers tend to use Accident and Emergency hospital facilities because they don't have a GP and because they don't have medical cards to move from place to place. When they do want to register there are forms to be filled out, going to Central Services, the forms come back but by that time the Travellers could have moved on. There is not, therefore, the preventative health care needed because there is no mechanism that facilitates a Traveller going from place to place. We need culturally sensitive equality. It is not equality if agencies are not achieving the same quality of outputs as for the non Travellers – the settled community, other racial groups.

Photographic exhibitions have been hosted in Craigavon and Derry to better highlight Traveller culture. Unfortunately resource constraints have limited the range of venues which could be tapped into in order to give wider exposure to the exhibition and, disappointingly, no residents groups or support bodies came forward to promote this work. As one interviewee put it:

> There were no community groups who would ever come up and ask about Traveller culture or request Traveller photographs to be included in their displays. It just does not happen that way. This is racism by neglect. These people are not seen as part of their cultural heritage. We just don't get requests from community based partnerships and organisations. The vast majority of people out there do not accept that Travellers are an ethnic group.

In relation to *farmers*, the following observations were offered on social exclusion:

> Farmers would feel that the rural development programme excluded them. It is hard to know if this is tinged with religious connotations since there is a perception that rural development at the outset was about getting Nationalists to engage. But farmers also felt excluded because of the whole community development approach. A farmer's ethos is the individual business, success, and profit for the farmer. The community development ethos is built around the community.

The perceived non inclusion of farmers in the first five years of the Rural Development Programme, which in turn gave support to community based projects of 'dubious sustainability', was highlighted, notwithstanding recognition that farmers are important "economic engines" in rural areas. Indeed the point was made that we should think of rural development as not having a capital R and a capital D since this tends to maintain a silo mentality with Government. The functional separation of Rural Enterprise Division (under the Chief Agricultural Officer) and Rural Development Division in the Department of Agriculture and Rural Development sends a message that the latter is less interested in the farming community.

The role of and opportunities for *women* in rural Northern Ireland were matters pursued during the interviews. The conventional perception of women as 'non-joiners' was encapsulated by the following observation from a membership organisation interviewee:

Women are told they can join farmers' unions, but when we ask anytime "Where are
you?" they are not there!

The view was expressed that while there are women who will never become
involved in groups, there are many who nevertheless seek access to opportunities
for social and personal development. The six women's networks across Northern
Ireland endeavour to provide support in this area, including the running of training
programmes. The following response, given to a question regarding the difference
between women's groups and community groups, highlights the broad constraints
and attitudes impacting in this arena:

Many women's groups are not happy touching economic projects. They want their
activities to be socially and educationally based. This goes back to how women
perceive their roles from their school days – they are pushed towards the arts. There
is the tradition that women are not involved in economic development. That is the
man's role! And thus we are scared to be involved, we are not used to the language
and we are not familiar with the processes involved. Secondly, there is the time
commitment required. The women come home from work and then take care of the
children. There may also be a care responsibility for parents and so on. So their time
is taken up and they are scared to commit to something that they will not be able to
run with. And thirdly the group formation thing also requires time. Somebody
coming in with a project to do will provoke the likely reaction "Aah – this is too quick
for me!" Now funders do not recognise this situation.

Within the farming sector it was pointed out that there is a big difficulty in
reaching out to women in farm households and getting them involved in women's
groups. A combination of limited time availability, accessibility constraints and
tradition condition the typical response "This is my place – this is my work". Indeed
many women's groups would acknowledge that they do not involve farm women as
fully as they would like to. While few women farmers are regarded as business
principals, their role in connection with book-keeping, tax returns, grant
applications etc is acknowledged as valuable and necessary, but frequently
"unsung". Women are perceived as being better communicators, perhaps reflecting
their training related to an off-the-farm career. This situation, which also has a
relational implication, was illustrated by one interviewee:

Usually the wife will come on the phone and describe the problem in great detail. And
then she says "But will you have a word with himself?"

The perception of women's groups by some civil servants was expressed with
stinging criticism:

The Minister is very supportive but the staff do not give us sufficient recognition.
They see women's organisations as 'bra-burners'. I think deep down they do not want
to admit to the value of women's groups.

Another illustration of rural social exclusion was offered during the interviews in relation to *sexual orientation:*

> We did some work with an organisation. This was trying to broaden its appeal since it would tend to attract a particular clientele. One of the related issues we tried to explore was male suicides in rural areas and the higher rate in rural areas and the possibility that this is linked to homosexual young men or women who don't have the support networks and don't have the opportunity to talk about it. But the organisation was not ready to take this on board. It backed off from this issue. It is not just about the willingness of service agencies to talk about these matters, the representative and sectoral groups are themselves very nervous about these things, perhaps coming from a very conservative background...We had a meeting to discuss our Equality Scheme with different sectoral groups that we are obliged to consult with. I went to the sexual preference workshop. This comprised an 'in your face' presentation. We were asked "How many people are gay in your organisation?" You could see people shrivelling in their chairs. A representative of a teachers' organisation was asked by the facilitator "How many gay teachers do you think there are? How many gay people are in your teaching organisation? Are you really saying to me that there are no gay teachers?" This was used to open up the whole discussion. I said that the support organisations for gay people don't exist in rural areas, this is not talked about and the situation is probably 50 years behind Belfast. So the issues around perception are more complicated for rural people than they are in the urban areas. Sexual preference is a difficult issue to quantify because it is very silent. People do not want to come out. But we don't know if the policies we implement are excluding them, because we don't know who they are.

In relation to those in rural society who are *elderly*, the following views were expressed:

> We are going to have to consider the aging population in rural areas as an economically active unit and, therefore, we are going to have to target the types of jobs, training and skills and so on, with the realisation that with labour shortages we are going to need older people to get back into work. Agencies tend to perceive older people as a welfare issue rather than as an economic activity issue. How we can help with re-skilling, for example, to get older people back into work is something perhaps that we have been insufficiently aware of in framing new programmes.

The issue of age was highlighted in regard to farmers where there is a varying perception as between Northern Ireland and the Republic of Ireland. In Northern Ireland it is more common to see a farmer in his 50s on the road driving an old tractor, whereas in the Republic of Ireland the industry is perceived as much more youthful and with up-to-date plant. Such an observation says more, perhaps, about the perceived uncertain long-term future of the agricultural sector in the region.

The position of *young people* in rural society has been closely examined by Youth Action – Northern Ireland (YANI). In 1997 it published a major qualitative and quantitative research study titled *A Sense of Belonging* which explores the experiences, problems and priorities of young people living in rural areas. It highlighted their interconnected experiences of being dis-empowered, under-valued and marginal to economic and social change in rural life. But it also recorded the complexity of these experiences and identified the ways in which young people were invested in for socio-cultural and sporting reasons. The report made a number of telling conclusions on community relations:

> While the issue of sectarianism tends to be raised more in interface areas, it is an issue that tends to be largely unspoken in rural areas. Nevertheless, if community development and regeneration are to occur, sectarianism needs to be tackled on an everyday, on-going basis whether or not direct contact is with 'the other side' (YANI, 1997, p.122).

However, the point is also made that religious and political differences are only one of the cleavages that confront young people in rural society:

> There is a hierarchy of difference operating within local communities and some differences are more acceptable than others. For instance, while differences in religious/political identity are more readily acknowledged, differences in relation to race, sexuality or ability may be less so. The manifestations of such intolerance are not always easily identified. For instance, people with a disability, while well catered for, may be dismissed or ignored, racist jokes may be a 'norm', whereas a known homosexual may be at the receiving end of physical assault (YANI, 1997, p.122).

In response to these findings Youth Action established a Rural Unit as one of five that the organisation operates. The initiative identified three distinctive needs among young people: (1) capacity building projects to allow young people to address their own needs independently and in collaboration with community groups; (2) testing methodologies with partnerships and community groups to encourage greater participation among young people; and (3) developing processes through which young people can influence policy making and decision takers. Three rural model projects were subsequently established to respond to different problematic conditions, comprising a largely single identity group in South Armagh, an interface area in South Tyrone and cross-border collaboration with youth services in the Republic of Ireland. Table 4.1 shows that these initiatives are still at an early stage but have achieved important preparatory and developmental outcomes.

The development of the model projects and the emergence of a distinctive strand of youth work in rural areas reflects the need to be proactive rather than reactive. A Youth Action manager emphasised this transition:

> In the past we have responded to existing initiatives, interests and personalities with an interest in this area. We need to move to a more proactive approach in the way we work and the areas we work in.

But the nature of working has also demanded a sensitive approach to programme design from which EDI could learn. In particular, this has involved a rational process of analysis and planning by YANI which ensures that actions are embedded

Table 4.1   Rural Model Projects by Youth Action

Project	Purpose	Activity
South Armagh Youth Initiative	This is an interagency response to address the social exclusion of young people aged between 14-25.	Throughout 1999 a range of capacity building programmes have been delivered to rural young people. Collaborative models between young people and community groups have been tested and young people have established a grant-giving group.
Rural Interface Initiative	This focuses on Clogher, Moygashel, Fivemiletown and surrounding areas; aims to enhance the capacity of young people in rural interface areas to address issues from living in contested areas and so work to transform relationships both within and across communities.	A baseline study has been carried out with young people aged 14-25 years in order to inform practice.
Cross-Border Action Group	This was an initiative in July 1999 to identify and confirm the key issues for young people through a cross-border regional support structure for youth participation.	Pilot projects have been developed in partnership with the National Youth Federation of Ireland, Louth, Monaghan and Donegal Youth Services. During 1999, 468 young people have been involved in these projects and 427 adults have contributed to the consultation and forward planning.

Source: YANI (1997)

in the organisational structures operating in a given community:

> We work to a distinctive methodology. It is essential for us to work with groups and
> networks operating in the area and not parachute in with all the answers. We identify
> the main gatekeepers in the area and attempt to work with and not for them. An early
> task is to define the needs of young people in the area and this helps us to refine the
> area or issues that we will work with. This can be anything that might work
> and, for instance, we have done excellent work with drama as a medium for young
> people. This can help them to voice their own concerns in a way that they are
> comfortable with.

Additionally, it is regarded as important for young people to be seen as a
resource to be used positively and to contribute to community renewal in rural areas:

> There are urgent priorities to develop the capacities of young people, promote their
> role as active citizens in rural society and recognise their leadership potential.

One of the most challenging areas of youth work has been with interface issues
and minority communities. As Table 4.1 demonstrates, Youth Action has developed
a multi-spatial approach to explore relational difference and distance in a number of
communities. But, as with EDI itself, this is acknowledged by YANI as having to
overcome a number of practical problems:

> There are huge problems associated with working in this area. Getting a venue that is
> genuinely neutral and accessible to both sides is a real difficulty in rural areas. There
> is a lot of resistance to this issue especially in single identity communities where there
> has not been any real experience of cross-community working.

Yet the issue of interface management is perceived by YANI as having
tremendous potential by engaging the problems and potential of young people in the
management of spatial conflict and contested spaces:

> There is enormous potential for young people to be a real catalyst for change in rural
> society. Our work on the interface projects shows that common problems around
> young people bind communities together much more than it divides them.

Culture and sporting activities represent arenas of activity with special appeal to
young people. Yet Youth Action are critical of the way in which power holders in
local rural societies dictate the activities, venues and opportunities for contact
among young people. Young people often have little freedom to negotiate
friendships and contacts outside the structural constraints which are reproduced in
local areas and which have built up for often political reasons:

> Many rural communities tend to be monopolised by activities and organisations that
> are perceived to be politicised, for example the GAA or marching bands. Often
> young people have only one choice – to join up or opt out of such activities. Even
> where young people do participate in such groupings the power tends to remain with
> the adults.

Developing strong internal social capital is a strong feature of self-sustaining communities. Infrastructure around sporting and cultural events can provide the basis for community organisation, networks of trust and reciprocity and physical capital in the form of halls and meeting places. But the points made by Youth Action provide important challenges as to how EDI is interpreted in the distinct socio-cultural landscape of rural Northern Ireland. Bonding social capital is a powerful dynamic in local communities, but it also has the potential to be exclusive, elitist and protectionist. The 1997 YANI Sense of Belonging report observes:

> Even diversity within a culture is not appreciated as often other sports/activities will not be tolerated or facilitated in local venues... Despite the 'non-sectarian, non-political' image which is attempted to be portrayed by the GAA, the majority of Protestant young people do not play Gaelic Games, nor are they encouraged to do so and they perceive the local management to be Catholic. (YANI, 1997, p.65)

Issues of social exclusion and EDI also have implications for those at the youngest end of the age spectrum. Playboard, for example, identifies strongly with the distinctive problems of *children* living in rural areas:

> We do see quite a vast difference between rural and urban areas in terms of transport, geographical isolation and economies of scale. We have convinced government to accept different rules for After School Provision so that you need 24 places for a facility in a town but only 10 places in rural areas. Rural isolated communities wouldn't support more places.

Playboard argues that a different approach is needed in rural areas. This should recognise that the support needs of rural children may not be in the sphere of fixed equipment and facilities:

> We need a couple of play workers in rural areas who can get out to isolated farm dwellings and villages. They may not need a playground which costs £140,000 for each facility. That amount of money would pay for four play development workers who could work in rural areas.

EDI is unlikely to have permeable effects without adequate services in rural areas which can bring children together on a sustained basis. Nevertheless, Playboard has attempted to develop community relations work in the play sector, among workers and especially among very young children. An interviewee from the organisation commented as follows:

> Playboard started to do some work on community relations because we witnessed sectarianism in children as young as two in a way that you didn't see elsewhere in the United Kingdom. If we give children the opportunity to play freely from sectarianism it could open their minds to other possibilities. The political conflict has closed people's lives in Northern Ireland and very young children can look at the colour of the paving stones and say whether they are in their own territory or not. If you are

playing under a peace wall it will never go away from your eyes or your mind... The silence, and the politeness, and the avoidance are done in a highly sophisticated way. We do it without even thinking and our children then learn how to do it too.

Issues of exclusion framed in terms of *religious difference* are identified in Northern Ireland as going beyond the conventional Catholic – Protestant, and larger churches – smaller churches dichotomies, to have relevance for other world faiths present within the region.   One of the key challenges to the effective and comprehensive implementation of EDI is perceived to be the uneven capacity of different ethno-religious and territorial communities to absorb its aims and associated delivery programmes. The introduction from the early 1990s of community relations training into many parts of County Antrim and County Down was identified as being difficult, though this has been changing somewhat in the past two to three years. The following observations were offered by one interviewee:

The Catholic community seems to have this greater sense of oneness; they are used to having to work together and to push things in their local areas. They have this ethos behind them. It may be related to the church or parish environment in that there is one local church that they go to. In Protestant areas, in terms of developing women's groups, people are very reluctant to go outside their area. For example we had an International Women's Day event last week in a neighbouring town and very few Protestant women turned up... There is a fear among Protestant women about asking for help. They are interested in product and do not really understand process. They are used to working on their own and thus looking for outside help is very much a new concept for them. While we try to include everyone in our programmes we still find that it is people from Catholic areas that are taking up the programme places.

Weak community infrastructure has long been recognised as a feature of depressed urban and rural Protestant areas. But this has been accentuated in rural areas where weak internal organisational and development skills are perceived to have left some Protestant communities impotent to respond to demographic, social and territorial erosion. This is most keenly felt by the Orange Order which has witnessed Lodge closures and mergers, especially within the border area close to the Republic of Ireland. As a community development manager attached to the Orange Order commented:

The Carlingford Lough District has had to amalgamate this year with Downpatrick. In the Downpatrick-Newry-Warrenpoint area we are having real problems with numbers and keeping Lodges open.

But the disintegration of some rural Protestant communities, especially in border areas, is connected to a range of factors that go beyond community organisational capacities.   Declining demography affects the institutional capacity of local communities to maintain a balanced community. The issue of rural school closures is thus vital, according to one interviewee connected to a cross community language group, in the protection and maintenance of a viable and vibrant population structure in beleaguered ethno-religious space:

> Country schools give an area its dynamism and identity ... without them a community
> has no chance and that goes double in Protestant areas. The state of community
> development in Protestant areas is desperate.

The symbolic significance of community places also has implications for their use and non use. With reference to women's group activities one interviewee commented:

> There seems to be a lack of awareness that they are running things in Orange Halls.
> And then they are wondering why people are not coming along. It has never occurred
> to them that this venue would exclude some people from being involved.

If the concepts of equity, diversity and interdependence are to translate to the social, ethnic and physical landscape of rural Northern Ireland, then its connectedness to the hard infrastructure (for example, community buildings, playing fields and monuments to the dead) and soft infrastructure (membership organisations, community workers and training programmes) that build community permanence and a sense of real and figurative place must be recognised. This is especially the case in terms of the producers and consumers of regeneration projects. A perceived major problem for the Protestant population has been its relative lack of success in negotiating with mainstream funding agencies and drawing down resources that are often vital to community reconstruction and physical regeneration. This is expressed by an Orange Order interviewee:

> We feel that there has been an imbalance in the allocation of funding over the past
> few years. All the blame has not fallen on the funders. Some of the blame falls on
> us for not engaging the funders.

The Orange Order has frequently articulated the problems within the Protestant community, and between the Protestant community and new governance structures after the imposition of Direct Rule in 1972. In particular, the distinctive cultural history of the Protestant people has been advanced as an explanation for part of this problem:

> The reason the Protestant community does not engage is partly historical. In one way
> Stormont was a disaster for the Protestant community in that there was a sense that,
> when Stormont was there, Protestants could get things done through the political
> establishment. As a result of this, community organisations did not grow and even
> declined.This is best shown in the Credit Union movement which is 30 years behind
> the Catholic community.

Moreover, the powerful socio-cultural, organisational and physical resources of the Orange Order have not been used as a basis of indigenous or single identity development. One argument in this context is that the orientation of the Orange

Order has evolved from a culture of defence and resistance:

> It is partly to do with our siege mentality. To understand the Ulster Protestant
> community you have to understand siege mentality, which is what drives it. It is a
> factor that influences almost everything that happens in the Protestant community.

Yet the Orange Order comprises a broad multi-level organisational network,
operating across the Northern Ireland social structure and in nearly all-Protestant
areas. It is perceived as being loaded with potential to define a progressive agenda
on community development, not just for the Order, but for Protestant groups in a
range of problematic locales.

> The dynamic within the Orange Order is at District level. The Lodge level is too low
> as they don't have the expertise at this level. But if you have 400 to 500 people
> involved, you get a sufficient bank of talent to have a dynamic.

The Orange Order has initiated a programme of community development which
has done much to establish a baseline of organisational capacities and to identity the
obstacles that stand in the way of the organisation becoming a real agent of change
for the Protestant community. In particular, it has recognised the historical problem
of weak development and under-funding in Protestant areas, together with the legal
problems encountered when engaging mainstream funding organisations. One
interviewee captured well the operational constraints imposed on the Orange Order:

> Funders have been aware of this problem and they have chosen not to address it.
> Some have difficulties with the Orange Order on grants for large capital amounts.
> But the Trustees who own the halls are not all members of the Order. Trustees can
> make applications but must also sign an undertaking that a hall will be available for
> letting among all bone fide community groups regardless of religion, sex, class or
> creed. This is creating tensions among members and legal problems for the Order.
> We have had to invent community groups that are manufactured to get the halls done
> up. So we need to recognise the reality of this situation.

But, legalistic and procedural problems are overlain by a range of cultural
subtleties that also prevent easy and effective utilisation of funding sources. A cross
community language group interviewee cogently argued:

> It is not the big things that are killing communities. It is the small things, like how to
> make an application or do a proper business plan. Some Protestant groups can't take
> money from the Lottery Commission because it comes from gambling. Surely a
> way can be found to take this money from another source and replace that pot with
> Lottery cash.

The need for small scale, basic level and patient support to Protestant groups was
a recurring theme and one echoed by the Orange Order in its attempts at local
development:

A problem is that we have an idea, make an inquiry and ask for information and then it runs into the ground. Our people need to be given a helping hand. Funders should recognise this. It is basic stuff.

The Orange Order, for example, plans to make an application to an unspecified source to upgrade the physical condition of around 30 Orange Halls in Co. Down in order to bring them up to entertainment licence standard. This, it argues, provides evidence of the level of need within Protestant areas and the hunger for support among the Orange Order. The scale of interest was summarised in the following way:

We ran a community development conference in Banbridge for County Down lodges to meet the funders. We expected to see 50 to 80 delegates but we had 210 replies and had to change the venue. In the end around 180 people turned up. There is a vast demand for this and we need 2 community development officers for County Down alone.

While the analysis above focuses on religious difference, it is worth reporting one illustration of intra-denominational tension which has promoted exclusion at the local level:

A group was starting up in the area, but it was saying "We do not want people from that area over there to join us". This was being said even though all the people in the area lived within a two mile radius of each other. Now we looked at this very closely and we discovered that it was not a Catholic / Protestant thing. It is about a priest who moved a church 50 years ago to a new site. The congregation had split and the enduring attitude was "Well, we do not speak to those people, we do not mix with them. They are there and we are here!"

But even within Protestant communities there can be relationship hurdles to be overcome in bringing forward social inclusion projects and as one interviewee outlined:

I told the church based group that it had to be representative of the wider community and that it would have to bring in other people. In other words I was getting them to realise that this meant talking to Presbyterians, Methodists and Baptists. Now that was as big a challenge for them to do as it would have been to start talking to Catholics who do not live in the village. Well they did bring in these other people and the interdenominational project will open soon.

The *Gaelic Athletic Association* (GAA) has played a vital role in the dynamic of rural life in Northern Ireland that reaches beyond sport to intersect with religion, identity, culture and increasingly, gender. There is a perception that identity and sense of place, and how the two combine to reproduce community solidarity, is still exceptionally strong in Nationalist areas. Nevertheless, the games and the organisation have continued to evolve in an increasingly secular world where the traditional bond between the parish unit and local people is regarded as playing a less important role than in the past. This is seen as reflecting, at a wider level, the weakening of religious obligation by more consumerist lifestyle behaviours and ambitions. But within this context the GAA perceives its role in and contribution to society more narrowly and with expectations that it will have nothing more than a single identity connectedness. The argument is that there is no problem with a strong sense of self-confidence built around socio-cultural and sporting traditions. This is held to be not threatening to others, while at the same time it is not compromised by other identities and value systems. Thus the need for a cross-community and cross-cultural agenda on every facet of life that stretches and even challenges individual choices and preferences is questioned by the organisation. The GAA recognises that this stance would not be universally popular. It has been critical of the poor media portrayal of its sports and the organisation itself, the sectarian attacks on players and club premises, and the low allocation of public funds, including EU Peace and Reconciliation funding resources. On the other hand in relation to inclusion, there is an explicit anti-racism policy in the games, participation by women in the games is increasing and, the policy debate on the eligibility status of members of the Northern Ireland and British security forces has formally ended. The GAA should be perceived as reflecting reality. It does not determine the structure of society and it is accepted that, to attempt to do so, would be wrong. In short, the organisation recognises the potent mix of conservatism and radicalism that dominates its discourses, which will have a significant bearing on how diversity and inclusion related processes will work themselves through, not least in rural areas. In short, rural communities are multi-layered: socially, spatially, economically and culturally. The tensions are more than religious difference and, as the comments above underline, the way that personal and organisational relationships are structured within a community or a locality can have major implications for participation and integration.

Social exclusion was also identified as a cross-cutting problem affecting a wide spectrum of vulnerable groups. One example cited was *homelessness* which could impact on Travellers, the mentally ill, ex-prisoners and sex offenders or those with alcohol or drug dependency. Interventions often require highly sensitive negotiations with local communities and collaborative work among a range of service organisations in order to secure temporary or longer stay accommodation including hostel provision. Within rural areas this more frequently involves the arranging of 'bed and breakfast' lodging. A frequent theme explored during many of the interviews was the need for service organisations to better promote the ways by which they do their work and to enhance the accessibility of this provision. This can involve the publication of materials in different formats (for example, large print, braille, audio tapes) and facilitating people being able to travel in to local offices from more remote rural areas. One interviewee highlighted, for example, the use of paid-for taxis to allow rural people to travel into a town for advice services.

Victims, survivors and ex-prisoners are some of the most marginalised and vulnerable groups in Northern Ireland rural society but they have been given particular attention in the emergent post-conflict setting. If organisational processes and problems of exclusion are to be embraced by EDI, then the reality of this deeply fractured and contradictory constituency must at some stage be confronted. This is especially the case in rural society where the networks and support systems do not appear to be as strong as in major urban areas. Programme managers operating in the victims field have noticed the spatial variation in the capacity and size of the victims' and survivors' sector as described by a support worker:

> In Belfast a network has been formed and there is also a support system in Derry. Most of my work is in a band right through mid-Ulster up to Coleraine and down as far as Newry.

Northern Ireland Voluntary Trust (NIVT) operates a range of initiatives that aim to increase and deepen the infrastructure but this has been especially problematic in rural areas for practical, spatial and cultural reasons. This has particular relevance for the scope of EDI, its method of application and the skills that will be needed to reach the hidden effects of violence:

> Victims and survivors groups may not approach Rural Community Network, for example, because they would be in a network with all sorts of other people and this lack of knowledge frightens some groups off. It can be a very risky place for them to be and they would be suspicious of who would be there and what their backgrounds are. Some networks are people with which they agree – we need to open them up to other positions and people with whom they have something in common with.

NIVT has found it difficult to engage some Protestant victims' groups that tend to be concentrated in rural areas and draw heavily on biblical teaching. Evangelical values are woven into a response to trauma that makes them resistant to conventional approaches to support and development:

> The main problem is reaching the highly politicised Protestant groups. These are very much rural based and class based. They would find the label 'community groups' degrading. They use their own counsellor, research worker and advisors and won't touch any other support groups. Funding them could perpetuate their isolation and they are not used to working in committees but rather they operate in hierarchical structures. They are often quite church-based which can be powerful and helpful, but it can often be difficult to see how these groups will progress.

The work is slow, highly politicised and controversial. The NIVT response has been to work to support the capacity and skills base of the community network but at the same time recognise that, especially in rural areas "people are so afraid of strangers and they need the capacity to handle things themselves" (Victims and Survivors Support Worker). But another vital obstacle to the development of this agenda has been the resistance among mainstream community groups to confront the issue:

> The community sector is very poorly equipped to deal with this. Most groups would never raise the effects of the Troubles and tend to stay well clear of it. People who

may have wanted to raise it were 'patted on the head and given a cup of tea' or were pushed to the side because they were perceived as a troublemaker. It is hard work trying to deal with these issues, it is very long term and it is connected to the political level. You can see why most groups will try to avoid it.

In the five years through to 2001 the number of victims groups in Northern Ireland has risen from 5 to 60 and this has been largely fuelled by the Peace and Reconciliation funding. So there is a danger that working with survivors as a distinct interest may simply reproduce their identity as victims and not engage a progressive movement in their attitude or behaviour:

> We don't want to sustain a victims sector. We want to help them articulate their needs but their pathway is maybe through other forms of local activity. We don't want to always see them as a victim but as part of a wider community.

Understanding of and commitment to EDI

EDI is recognised as becoming a valuable and real language to break down what is meant by community relations since it allows people to make a whole web of links in their work based and personal relationships. More particularly it allows conversations about the "s" word (i.e. sectarianism) to take place. It is interesting to note the views of one interviewee, who eloquently captured the contemporary timeliness of EDI:

> I think there is the beginning of the search for a new conceptualisation, new patterns of praxis, it is about catching a glimpse of a vision.

EDI was defined in various interviews as being "good manners", "awareness and trust" and "citizenship". EDI thus becomes a part of the way we live our lives within work and outside work. The language was criticised on a number of occasions as being "off-putting jargon". Within a community setting it was suggested that the following style could be used to introduce the concepts:

> I would ask "Can we reflect about what is going on in our community? How can we get more people involved? Are we truly representative of all the interests in our community? What do we think of people down the road?" And so on. It is a language like this that we have to use.

EDI was cited on a number of occasions as being more than community relations which was described as comprising "grants to the arts and the local pantomime, hampers to senior citizens at Christmas, a fireworks display, or poetry competitions on themes of peace and reconciliation between schools". The EDI challenge is to shift the focus from what a Community Relations Officer does to what an organisation should do. EDI is certainly not about finding the token Catholics or the token Protestants for cross-community projects which was regarded by one interviewee as "very manipulative". In this regard one opinion was stridently expressed that local government, and in particular its elected representatives, must become models of support for EDI dispositions:

> It must demonstrate that it is fair to everybody and that everybody is valued. It is really advancing a type of citizenship, a type of civic leadership which is working for all the people and is able to face up to the particular context it is working in. To start that process the people who have political leadership must be on board, the people who can make it happen, who can give the authority, space, resources and blessing to allow other members of the organisation to follow suit. If the leaders don't do it then nobody else is going to value it. So it has got to start with the political representatives. They have got to go through their own educational process and their own training; they have got to come up with their own set of combined values, rather than separate values which despise other representatives and dictate how they treat others. The organisation has to move away from being a place for pit bulls to snarl!

Newry and Mourne District Council has been active on this front. One of the first things it did was to devise and publish a voluntary code of conduct which for lays down principles and values. This explains how Members should relate to people who are completely different and opposed ideologically to them in public, within the Council chamber and in the Committee room. The important point here is that this initiative was operationalised at a senior level. Moreover, using external expertise there was a gradual transference of the necessary skills and confidence to work in this way. There is, however, an appreciation that this necessary step change in organisational culture is dependent on a continuous effort to sustain learning capacity. One interviewee also questioned the easy acceptance of EDI as another perceived layer of checks within the public sector. The view was that there will be reluctance to formalise the adoption of the process unless it is made mandatory like Equality Schemes. On the other hand, the point was made that EDI could have major benefits for the voluntary sector which tends to work within "an ethos framework" compared to government organisations which operate within a legislative framework. As one interviewee suggested:

> EDI is a new way to start to analyse what we are doing; to ask ourselves "Is this what we are about?" To look at the kind of messages we are putting out; To look at the training programmes we offer rural communities; it is a self scrutiny of values.

Many interviewees sought to explore the meaning of EDI terms and to relate these to their particular concerns. Thus the Travellers' Movement argues that equity revolves around equality issues, but that equality for Travellers should be equality of provision in a culturally sensitive way:

> when Travellers do not have to change their lives and their culture to get the service provision to which they are entitled.

Interdependence for Travellers must be constructed around concepts of empowerment and independence:

> In Northern Ireland Travellers tend not to go to secondary school, they have an extremely low literacy and numeracy rate, and they don't go on to third level education. Travellers are starting from a lower base, because of education, compared with other groups. Therefore, we don't have the Travellers with the skills to take on organisational posts. We need to give them the skills so that they can have independence. When Travellers have been through a community development process they will begin to act on their own behalf. It will then be easier for them to act in an interdependent way. Interdependence is the next stage *after* independence. Only then can Travellers really begin to interact effectively with other organisations.

In discussing social exclusion and division it was interesting to record how organisations were themselves perceived by some rural constituencies and how the organisational reaction to that perception was framed within the context of an EDI perspective. One interviewee commented at length on this matter:

> I got a fairly direct letter which said that commissioned research indicated that my organisation is perceived as a sectarian organisation, that it is biased against other organisations coming from within the Protestant ethos. We then had a whole day meeting and the conclusion was that those groups did feel that we were operating an anti Orange Halls policy. The core issue for me as I was leaving is the need for us to decide what is our community relations policy. I was beginning to hear things that I would never have heard before. If you don't hear them then you are not going to be able to adjust what you are doing. That is a healthy environment to have...Now the important thing is that if our organisation is perceived in a certain way that is a valid perception that these people have. There is a lot of soul searching to be done by our organisation and others like us. It is necessary to get our community relations policy in order. It is about building it around EDI. Our programmes, eligibility criteria and targeting priorities have to be within a framework of community relations or EDI related to the whole organisation. We need to sort out internally whether we are going to target single identity groups or cross community groups or whether there should be a baseline below which we will not go in terms of delivery. It is naive to seek to implement a rural development programme which is focussing on partnerships between both sides of the community – the motherhood and apple pie version of EDI that we all love each other and we all get along. I am sorry it is not like that out there! We are dealing with very bitter sectarian groups from both sides of this community whether we like it or not and we are going to have work with that situation. We are happy to work with single identity groups but they must compete along with everybody else in terms of the quality of their projects. Just because you are the GAA or the Orange Order will not get you a further rung up the ladder. If you have a good project and a problem to be solved then we will work with you. The baseline below which we will not go is if an organisation is building bridges away from the other side, if the support helps dig it into a trench, if the support pulls the shutters down. On the other hand if in some small way the assistance is leading to convergence, even if it is ten years down the line, then we are interested. This does not mean that any organisation has to give up its culture whether Catholic or Protestant, Nationalist or Unionist. There has to be bridge building across, no matter how tentative. That is the

sort of EDI framework that we need to get staff to buy into and our board of management. But I know that there would not be enormous support for that approach. I know that people would have entrenched views and would say that we should not focus on groups who are not prepared to work together. The political landscape out there is not like that. I believe that the cross community activities are not going to succeed unless you work with the groups who are not cross community. This is a key policy decision for us as an organisation and it is a debate that we still have to conduct internally.

There was frequent mention that EDI is a long term project which has to be gently nurtured. As two interviewees put it:

You cannot do this in 'chunks' which seems to be the way that the Equality Schemes are being brought forward. You cannot click your fingers and it happens. It is an organic thing. It is about gradual change. To see where it is going you have to look at the trends over time. You must avoid any snapshot appraisal.

EDI must permeate all parts of an organisation. It is not a quick solution and it is not a policy mixed in a day. It certainly is not about ticking boxes. It is a live process and it is for life. Thus there do need to be review systems to allow for follow on work within an organisation. Somebody in the organisation has to be designated the EDI leader.

In short EDI is regarded as facilitating a step change in value systems, from being non-sectarian to becoming anti-sectarian.

Equality, New Targeting Social Need imperatives and the EDI implications

An important challenge to the conceptual base, application and understanding of EDI is its connection with policies and practice resulting from the equality legislation, human rights and Targeting Social Need policy environments. The need for clarity is important because of the possibility of mixed messages being picked up from the 'noise' around these related agendas, the need to maximise the effect of an EDI programme, and the possibility that some agencies are preparing, or are likely to prepare, equality schemes and conduct impact assessment and policy proofing. This is also important because there is the strongly held view from within the Equality Commission (ECNI) that equality law in Northern Ireland is primarily aimed at the sort of systemic change implied in EDI methodology. In this vein ECNI makes the point that there are important differences between the Great Britain and Northern Ireland approaches:

There are things in the Commission for Racial Equality legislation that we would like. Where we score over the GB legislation is that ours is integrated and that it is positioned at the point of departure of policy development. You have more influence at the start of the process rather than have to chase people through the courts afterwards.

A policy that has an 'upstream' emphasis is designed to assemble the values, organisational structures and system changes necessary to effect a cultural shift in equality:

> We would not want to give up strong front-end policy but to do it properly is time consuming for ECNI, for public bodies, for the voluntary and community sectors and the affected groups. One of the reasons that it is so time consuming is that we do not have a sufficient practice base and enough related skills. What ECNI is trying to do is to get this to become second nature to people and not to say "Oh, we have jumped through the hoops and that is that". It is going to be very overt for the first couple of years until it becomes living, breathing, second nature. By that stage organisations will have created relationships so that they are getting the policies right from the start.

However, the way in which EDI relates to 'downstream' policy outputs must be clearly articulated in the rural policy sector. The ability of the concepts to protect against the unintended negative effects of programmes and decisions needs to be linked to the Equality Commission's attempts to develop Equality Impact Assessments as a tool of public administration:

> Equality Impact Assessment guidelines, especially linked to policies coming out of the Programme For Government and the budget, represent an important instrument for departments and agencies. This is a pilot stage and we will be undergoing consultation at the same time. If the guidelines are not strong enough or have not been applied sufficiently then we will look to see whether they need to be incorporated into statutory guidelines.

The Equality Commission emphasises that the laws, policies and systems now in place are about restructuring not just organisational attitudes, but also the fundamental link between government and the people it plans for:

> The Commission for Racial Equality is about making public authorities do things and going after them in the courts if they don't. The Equality Commission is about changing the fundamental relationship between policy makers and people in society. The agenda is for everyone, at every level.

But the Equality Commission also sees important preparatory tasks needed among agencies not yet designated under the equality legislation, but which are likely to be when the main government departments have had their schemes approved. This is likely to have effects on voluntary sector bodies and organisations supported by public funding or given intermediary status under EU or mainstream programmes:

> Designation has left out many bodies in Northern Ireland, including Intermediary Funding Bodies which dispense government resources, and we need to see some designation gaps closed. The ECNI has a long list of bodies that it would like to see designated under the law.

The Equality Commission recognises that, at present, the good relations duty plays a less prominent role in the formal Equality Scheme provisions. An interviewee from this organisation expressed important distinctions between equality and community relations in a Northern Ireland context in the following way:

> The Equality Scheme focuses on the equality duty but not on the good relations duty. The equality duty takes precedence over the good relations duty. You will never build good relations on inequality. We want to see good relations, not just good community relations and the former is more than or wider than community relations. EDI fits very well into that. We have stated the type of society that we are aiming for as being a truly participative, cohesive, diverse and respectful society. Adding good relations into equality is adding depth to the purpose of equality.

It is also important to recognise that the reach of the equality law goes well beyond community relations in the conventional sense of the term. The implication is that, for EDI to have maximum effect, it must also apply to other divisions and excluded groups in Northern Ireland. Tackling the formidable array of socio-spatial cleavages in rural society presents a significant challenge and emphasises the need to target limited resources effectively across a wide span of issues. The Equality Commission, however, underscores the need for an integrated and holistic response to injustice:

> There is no reason why one issue should be more important than others. On the Structural Funds Monitoring Committee we have EU officials who are more aware of the European perspective especially on gender and not aware of the Northern Ireland position. We cannot only concentrate on women and so the Department of Finance and Personnel must negotiate on the other issues as well. We would expect to see the same rules applied to all the other areas of the Northern Ireland equality agenda.

The interviews included a question on the attitudes of organisations to the preparation and implementation of these Equality Schemes. One interviewee suggested that these were "full of civil service language that we do not understand". Another interviewee suggested that Equality Schemes had remained within Government department headquarters and had not filtered down in any meaningful way to street level practitioners. There is, however, some appreciation that the preparation of Equality Schemes has forced new conversations about exclusion compared to the previous Policy Appraisal and Fair Treatment (PAFT) initiative. This view was expressed by one interviewee as follows:

> There is a hierarchy of attitudes to exclusion. There are the groups that you tick off and the processes you use to get to that point become accepted practice. Section 75 legislation has facilitated those conversations about marginalised groups such as Travellers, ethnic minorities, people with disability, age structure and so on. It is against this greater confidence that people can then move on to talk about sexual orientation with an ease, whereas initially the reaction would have been that this topic is not for here.

But the additional workload connected to Equality Schemes received frequent mention and as one interviewee observed:

> The whole equality agenda is generating so much stuff on the desk. The pile just gets higher. I don't think the processes will make any qualitative difference for our organisation, except that the Equality Scheme has forced us to go and talk to all these groups. This is the key difference – it obliges us to identify who we should be consulting with; it obliges us to take time out to consult with them. This is very important. But I don't know if the statutory rigour of the whole system is adding any real value. Rural development was set up to target social need – so we were well down that road anyway, more so than other organisations. But it will tighten us up. Another thing is not a lot of advice was coming from the Equality Commission. We missed the first deadline like a lot of other agencies, but we were struggling with it. It is only now that the Equality Commission has moved into a training mode to help agencies and bring them along. If the Equality Commission is going to behave like 'Big Brother' and slap you without giving you assistance to demonstrate how you can do it, then it is not going to work. They have now brought the reference groups together with the delivery agents. But the consultation groups are reaching burn-out as well. There were a lot of standard comments back on our draft Equality Scheme which could have been against any Equality Scheme – sort of process stuff. But we did get good comment back from the Rural Community Network. They know our constituency and what we are about. They made suggestions about how to do things. But we didn't get comments, for example, from interests such as the Orange Order.

Most significantly that interviewee added:

> EDI could become more important than the Equality Schemes. If there is an in-built, self-moving EDI policy and process, then an organisation should not fall foul of Equality and New TSN. This will be a horizontal thing that will drive all these vertical programmes. If it is a check across the whole organisation then it should be welcomed. You will never drive a statutory Equality Scheme through if you don't have an EDI appreciation at the outset. They are, therefore, closely interdependent. It can inform a whole range of activities within the organisation and how it interacts outwards.

Within Northern Ireland community-led local development has been pursued as a prominent approach to the stimulation of cross-community activity in areas of socio-economic disadvantage. The extent to which this conventional approach to collective action can underpin systemic EDI change was challenged by one interviewee:

> Take, for example, a cross-community project that is doing wonderful things for everybody – this will be perceived as the community relations content of economic regeneration. People will say: "isn't this wonderful because we are demonstrating that Protestants and Catholics, Unionists and Nationalists are working together for the common good by developing an economic project in our town". But of course the Protestants and Catholics are self selecting themselves to be together. We then have

everyone gathered and an important person comes down and cuts the ribbon to open the project and everyone claps and that is what is passed off as community relations! We really do need something radically different, that takes account of diversity, the divisions, tensions and inequities around, which includes the people who are being left out of that development process or who perhaps have excluded themselves. Such an initiative must demonstrate sensitivity to the totality of local diversity; it must be about binding people together in a new way not done before.

The view was expressed that there does need to be an EDI framework established for each organisation in a way which allows the principles and their operationalisation to connect to Equality Schemes and New TSN. Indeed it is argued that if an organisation embarks on an EDI process, then it will end up with an Equality Scheme which is watermarked by genuine appreciation of organisational context. Two interviewees expressed such sentiments quite forcefully:

> It gets us away from a narrow PAFT proofing exercise which is about ticking boxes, but it is more complex than putting equal amounts of money into each electoral ward or addressing imbalances by skewing resources. EDI in operational terms is about education, training and cultural change.

We need EDI because it is the difference between following the letter of the law and actually facilitating social change in the wider community. Some of the draft equality schemes have been very much tick box exercises; they have been very correct in what they have promised to do, but there is not much sense of real commitment behind them. EDI allows us to go much further and much deeper than that. It changes the culture of organisations, how they work.... When you look at what is out there, you need a mechanism to translate this stuff; Equality Schemes come in by the bunch – they do create a culture in their own right and that is very positive, but even so the degree to which they emerge out of a legal requirement to a point that you believe that this is something to commit to and frame your organisation around – that is a long process. You need mechanisms to find a way through for each other and EDI is one such mechanism. The equality legislation will do certain things – it will make you wake up in case you are infringing in certain ways – but does it act as a transformational tool?

Interviewees identified a number of situations where EDI principles have relevance for public policy. For example, the impact of physical planning policy on Travellers was illustrated with the following comments:

> A small six pitches site in Craigavon went to a public inquiry which took two years to report back. Now I don't know what public inquiries normally take, but surely they don't take 2 years! It is very difficult for Travelling families to get planning permission to build their own houses or to get approved sites for their family unit where they can put in a septic tank. We need a code of good practice on how Planning Service will work with other public bodies on Travellers' issues.

One issue with particular relevance for rural development is the perceived uncertain relationship between the preparation and implementation of Equality Schemes and 'rural proofing'. The view was expressed that urban-rural differentials are not part of the equality agenda and that space affects everybody, whether urban or rural. Yet it was conceded that, under rural proofing criteria, questions can arise around equitable service provision for different client groups for whom local geographies really matter. In the next EU funding round through to 2006 the point was made that in line with New TSN the rural development programme will seek to target some 65% of funding on disadvantage, especially group disadvantage among farmers and farm families, young people, rural women and the long term unemployed. Eligible areas are located outside Belfast and Derry and those towns with a population of more than 5000. It is within this context that the comments of one interviewee are interesting:

> Rural proofing could be part of the equality agenda. Rural geography and in particular, rural isolation, need to be looked at and, perhaps, put into the legislation. The apparent confusion between rural proofing and equality requirements is an issue which needs to be tidied up. I think there is validity in saying that if you are disabled and living in the remote countryside, you are worse off than somebody who is disabled and living within a town. We all need to look at this.

On the other hand, it was suggested that the formulation and implementation of an organisational rural strategy could be located under the wider adoption of EDI values thus helping to ensure that a necessary rural perspective was being supported. This point was underscored by indicating the need for sensitive local scale information, obtained in part from rural baselining, to help inform the EDI stance. The difficulty of relating the Sub Regional Rural Support Network areas (characterised by new geographical and straight line boundaries in some instances) to administrative boundaries, which are meaningful to public bodies, was cited as a constraint in this regard.

Training in prejudice awareness and reduction

Training in prejudice awareness and reduction was frequently interpreted as having relevance for the effectiveness of an organisation in dealing directly with its customers. There was common appreciation for the importance of fairness and organisations often claimed the absence of any proven discrimination as an impressive track record on anti-sectarian behaviour. This external consciousness tended, however, to mask a responsiveness to the need for critical self-reflection on issues such as sectarianism at the intra organisational level. Thus, for example, a public sector interviewee commented as follows:

> Our equality training was really specific to our Equality Scheme and New TSN. We certainly did not look at conflict resolution or sectarianism, certainly not during my

time in the organisation. These subjects have never really been discussed in a formal way. We do have general circulars on harassment, sectarian comments, the flying of flags and emblems. But certainly nothing specific. I don't know if it is because it has never arisen within our structures here, but certainly it has never been brought to our attention...

This interview was especially interesting because it involved three people from that organisation, with the person solicited for interview having delegated the task to these colleagues. Regarding the above comments, this spokesperson, in our view, was clearly perceived by colleagues as having difficulty with the question as is illustrated by the authoritative interjection at that point from another person around the table who said:

There is a general awareness that public officials being public officials are supposed to be totally objective and fair. Underlying all our dealings with staff, the public, with different groupings, with the community sector, we are obliged to take that neutral stance and to ensure that our organisation is reflected in its best light. We are not totally immersed in it, but we are certainly aware of it.

The initial spokesperson then rejoined the conversation and added in support of that statement:

We have all had to go on an equal opportunities course!

It is striking, therefore, that the Community Relations Council (CRC) should make the point that equity, diversity and interdependence go well beyond the equality agenda by helping to transform organisational systems and procedures in a meaningful and lasting way. But the biggest challenge to this task, it argues, will be within the institutions and structures of government departments and agencies. The CRC analysis suggests that rural conflict, interfaces and enclaves are not constructed as mainstream concerns of the relevant policy agencies:

We have, what Scott Bollens calls 'colour blindness', among the main statutory planning bodies. 'What ever you say, say nothing' is the policy response and many people in government deny the relevance of these problems to rural development programmes.

The CRC argument is that Government in Northern Ireland has avoided or denied contentious issues such as territoriality and spatial segregation in order to de-politicise policies and avoid controversy. Technical rules and decision-making processes have helped to insulate professionals from the sectarian tensions that often arise in resource allocation decisions. This has protected government agencies from accusations of discrimination and bias but as CRC points out, this can have costs:

Not having a policy does not mean that there are no negative effects. Communities can be left on their knees because their local school couldn't stay open or their post office closed.

The Council argues that a proactive policy is needed to ensure that programmes and policies are sensitive to the conditions of a divided society. In particular, EDI offers a framework to allow Departments to move beyond their narrow technical remit and to engage the substantive effects of their decisions on single identity communities, areas of social need or the prospects for developing integrative solutions to local problems:

> People need to openly address the effects of doing business, planning resources and working together in a divided society. We need ways and methods of learning to speak to each other in a positive way about how to get the best from limited resources ... We have a situation at further education college where students can't get to it or feel safe when they arrive because of where it is. The local railway station is accessible to both religions and so where we put facilities matters to people's use of them. It makes common sense.

EDI thus contains the methods, training insights and basic vocabulary to sensitise policy makers to these issues and subtleties, particularly in those situations which have an underlying tension or prejudice. The observations of one interviewee underline the potential of a programmatic approach to EDI transformation:

> There is a need to consider carefully how a confrontational issue can be addressed. Perhaps it is not better to take it head on, but rather to consider how best the issue can be brought in through a programme so that it enters the culture and in that way you can begin to address the issue. If it enters through a programme there is a real corrective process that goes on. It does have an effect on the organisation and its people in that they cannot go on making those remarks.

Indeed the relevance of training for building mutual respect as a response to perceived institutional insensitivity was illustrated by the Travellers' Movement:

> I went to Planning Service to talk about its Equality Scheme and Area Plans. There was no problem with this because there is no point in the planners going straight to the Travellers to talk about these issues. But when they sent me a letter back I was so irritated because Travellers was spelt with a small t. That to me is institutionalised racism because the planners do not see Travellers as an ethnic group. Whilst we had a very good discussion on the day – you wouldn't write to a Chinese person with a small c; what is important here is where people put you as a group and if you are putting Travellers with a small 't' you are not giving them the respect due.

A number of interviewees confirmed that training needs related to EDI are being addressed. NIVT has developed a four staged training programme aimed at existing community groups, residents associations and victims organisations. The REAL (Recognition, Empowerment, Awareness and Learning) programme operates in four sequential modules:

Section 1   Culture history and politics
Section 2   The effect of trauma on individuals, communities and societies

Section 3   Group processes, conflict and problem solving
Section 4   Organisational systems, managing and financing groups
             structures

The training has a heavy experiential emphasis in line with other approaches to community relations and conflict and has a strong rural inflection given the spatial gaps in support coverage in Northern Ireland. A significant point is that 20 public sector officials are going through the training and the programme will bring this cohort into contact with 40 participants drawn from the community sector to compare experiences and draw out common lessons and implications for future policy and practice. The Travellers' Movement has produced an anti-discriminatory guidelines pack which is Traveller specific and has to be delivered by a Traveller to Travellers or between a Traveller and a settled person. The way in which training and educational programmes should be devised was broadly summarised by one interviewee:

> The content should reflect what local people meeting together over a period of time have come up with. It is grassroots, it is locally honed, it is bottom-up, it is contextual, it is dealing with issues which are not the same in other parts of Northern Ireland. This gentle way in allows the confidence levels to be raised and the trust relationships to be built. In terms of a religion focus, for example, it is not just about bridging Catholic – Protestant issues, but also addressing the appalling levels of ignorance between the various Protestant traditions. Out of a process of trying to understand each other's traditions, the dynamic can go up a gear, people can perhaps go on to look at issues of violence and a culture of violence, of building community, of dealing with sectarianism. But this is a slow process which can take a number of years.

On the other hand the comments of one interviewee (who had attended a prejudice reduction training programme organised by a District Council) signpost some of the practical limitations which can be encountered in this type of capacity building work:

> The Council employed a big consultancy firm from Belfast to do the facilitation. But when we turned up and saw who was there I said "We are all workers, we are not the people you want here. You need people, for example, who are involved as volunteers in their local residents' associations". The information and invitations had gone out to the wrong people. Basically the Council had selected those people with whom it is traditionally involved. So at the end of the day they really only have a very tiny insight into what is actually going on. Then I asked the question "Where are all the councillors? Where are the Council officers?" There were none attending the training course. It was also targeted at the churches, but only one clergyman came and he had to leave early. It was all just workers! They had the time to attend. I ask you "How are people involved in residents' associations going to attend an all day training?" The training has to be time friendly. And there has to be an emphasis on real relationship building.

Playboard has also been involved in comparable work. Its first training initiative *Play Without Frontiers* was concerned with how to introduce cultural difference within the play environment. More recently it has moved to a *Games Not Names* approach that enables play facilitators to talk about their own sectarian experiences in order to develop models of practice for ensuring that these did not spill over to a future generation of children. The importance of starting slowly and incrementally and in single identity settings were also stressed by Playboard:

> We were very nervous about this and started this initially in a single identity work setting. Initially people were angry about this because some workers felt it showed a lack for trust. But as some highly personal and sensitive experiences unfolded, most of those involved accepted that the more incremental approach was more likely to succeed rather than getting into a cross-community situation "where there could have been World War 3".

Games Not Names is being developed and targeted at youth workers, play workers and teachers in mainstream education. It analyses the problems that children experience through living in a divided society and describes the techniques, training approaches and games that can be used to develop anti-discriminatory and prejudice reduction practice. Some of the best practice principles are:

- promote prejudice reduction in all play schemes;
- provide positive play opportunities;
- develop anti-discriminatory practice in children's and leader behaviour;
- develop linkages with other agencies with relevant skills and insights;
- promote children's understanding of their own and other's cultures;
- deal positively with sectarian incidence;
- leaders should develop mediation skills in managing sectarian disputes among children;
- make contacts with other play schemes adopting community relations strategies;
- ensure that venues for play work are welcoming and non-threatening.

A key challenge in the whole approach has been getting key stakeholders in the statutory sector, children's organisation and some communities to recognise that the issue of sectarianism forms a central and legitimate part of the play environment. Overcoming significant cultural attitudes such as these will be a vital challenge if EDI is to permeate the rural policy arena. Accordingly, Playboard makes the point that its training emphasis starts with an approach based on problem recognition:

> Giving people space to recognise that sectarianism is a major problem in the play environment is an important first step. We don't have answers but it is about getting the issue recognised and showing that play can be very powerful in community relations.

Playboard nevertheless recognises that there are dangers by concentrating exclusively on religious differences and that other important cleavages are opening up in the play environment with particular relevance for children:

> We don't just deal with sectarian matters. We have carried out research and training on both gender and disability issues. We published a document called *Gender Matters* which is about how to avoid stereotyping boy and girl forms of play.

The *Gender Matters* research makes the fundamental point that, "Playgrounds and play work practice can operate to reflect and reinforce sex role stereotypes. Alternatively, they can be a means of identifying and challenging those stereotypes, liberating each child to explore the full range of their potential through play" (Playboard (1994) Gender matters: A Guide to Gender issues and Children's Play, Belfast. Playboard p.7). The approach developed by Playboard is based on training in non-discriminatory selection methods for recruitment panels, equal opportunities training and gender awareness training for staff and volunteers. It also calls for an open approach to play structures and materials that can help to avoid the stereotyping messages conveyed in some commercial products. The point is made simply, but effectively:

> It is not right if, when working on the duty to promote equality regarding age, we don't even ask them what they need. You would not do that with adults and you should not do it with kids!

Indeed this core relationship between skills, training needs and EDI is reflected by the establishment of the Community Relations Training and Learning Consortium (CRTLC) some 18 months ago in response to the growing diverse needs of the sector and the withdrawal of the Community Relations Council from formal training provision. CRTLC has a number of objectives relevant to the application of EDI in a rural policy and problem context. These include the co-ordination of existing community relations training and the development of a classification system as the basis for longer term planning and development. In particular, it is recognised that there are differing needs among different areas and groups and that a clear progression of support is required within the sector. Up until the establishment of the CRTLC, training was delivered in an ad-hoc fashion. There is a perceived need to provide a clearer map of the skills, courses and methods available. CRTLC also aims to facilitate collaboration so that people can tender for different types of contracts on the basis of joint applications. The Consortium has been establishing a quality framework to maintain and build standards. This acts as a baseline to reflect on practice whereby trainers are encouraged to be mindful of the context of their work and the relevance of their programmes, and to constantly question their own practice. The Consortium does not have a large rural membership and as a CRTLC interviewee put it:

> Membership is Belfast based, but that is something that we would need to rectify.

CRTLC also promotes accreditation where relevant and draws especially on the Open College Network But it feels that new and innovative ways must be brought forward for developing skills in community relations training and education if concepts such as EDI are to have any prospect of success. The need for change has, in particular, been accelerated by the equality agenda:

> The equality agenda has created a monster for the Consortium. We have been bombarded with requests especially with regard to the promotion of good relations. People want to know how to move beyond a position where they are just ticking the boxes to a place where their organisations are totally committed to equality. It is about moving along a continuum from compliance to commitment. People are looking for a road map through this, a training manual that will help them develop good approaches.

This move from compliance to commitment will involve a formidable transitional change for which, it is argued, key service organisations are poorly prepared, ill-trained and culturally not attuned, especially with regard to the need for a proactive agenda in mainstream policy areas. CRTLC makes the point that this is vital in the transformation of local society along with the value bases and perspectives of those who design and implement policies:

> Implementing good relations is a huge challenge and we are in the process of developing practice on what good relations would look like in an organisation and how we can humanise it.

But it is accepted that the process must be an evolutionary one and cannot be imposed on a structure, strategy or organisational culture that is simply not ready for it:

> The process is about opening up spaces for people in organisations to have conversations that they would not normally have. Some feedback we received indicated worries that 'this is going to cause a lot of trouble'. It is not about that. It is about creating the conditions that allow people to express their curiosity and express freely about how they feel. It cannot be forced.

CRTLC strongly believes that this is where EDI has a clear role to play. In particular, it can work to facilitate a change management process that is systemic rather than partial, is proactive and not reactive and that reaches the hidden routines embodied within large bureaucracies that have stifled progress towards good community relations planning. But CRTLC also makes the point that the concept is not well understood. The clear implication here is that the concept needs to be articulated in terms of its relationship with the equality agenda, human rights legislation and New Targeting Social Need:

> EDI is not applied properly by some organisations. They don't understand it... Interdependence is the most difficult element to understand. It encapsulates equity and human rights but is poorly appreciated as a non-threatening way of dealing with diversity in the workplace.

One of the perceived obstacles to the effective development of an EDI agenda is the degree of competition between the community development and community relations sectors in terms of their perspectives and methods of working. It is also important to recognise that self-protection is not the sole preserve of central policy makers since a range of voluntary sector organisations and support programmes compete for scarce resources and seek to claim attention from policy makers:

> It has been difficult to establish links between the community development sector and it goes back to the competition between sectors and suspicion of community relations in particular. There is certainly resistance among both sectors. There is uncertainty about where the interface is and different analyses of what the two are. A lot of it has to do with funding. We need to strip off old labels and incorporate community development, community relations, human rights and equality under a common approach.

Moreover, there are concerns about the capacity of the training sector to fully absorb EDI across a policy field and thus, in particular, the needs of trainers cannot be overlooked:

> There is not enough space locally for trainers to up-grade their skills. They have to go to the United States. But we must seek to develop and sustain our indigenous skills.

The organisation feels, therefore, that there needs to be further investment in the range of skills that are currently available thus giving support to the infrastructure within the sector. CRTLC is attempting to build a stronger network of trainers by linking them in a more strategic way so that a different combination of methods can be applied in response to local needs and circumstances. EDI training has an important place in that configuration.

Dialogue with and among rural people

Conversation can be regarded as an essential requisite for the promotion of inclusion in rural society and yet there is the paradox that the more local these conversations become, the more difficult it is to confront exclusion and sectarianism. At the local level it can be difficult, for example, to get the desired diversity of participation in community based organisations. The challenges for rural communities in this regard and the connection with EDI prompted this reflection from an interviewee:

> When you are living in a local rural community and you want to say that we stand for the inclusion of Travellers, we stand for the inclusion of ethnic minorities, those with disability, gay and lesbian people and so on, – well you can't just utter those sentiments! They have huge consequences in terms of how they are heard. The challenge really is for rural communities to lead on these issues. Looking back I ask

the questions: "When in terms of social justice, and in terms of equity and diversity have rural communities taken the lead in society? Or do we always work in the aftermath of it becoming politically correct in the mainstream?" It is almost as if we are accepting that rural communities through their conservatism don't lead on progressive governance and wider issues. Now that is acceptance of policies being urban driven. If we continue to allow them to be urban driven that change is not going to happen. The challenge is to point out to ourselves where we are taking the lead on progressive governance thus showing to society at large that we are not afraid of big issues. But in all this it is important to appreciate, without romanticising it, that conservatism has a richness, there is tradition, relationships are often deeper, and care may be of a better quality. We should value those things that may not be present in the urban setting. The challenge in short is to say "What is the reality here? How can we build on that reality? Are we always going to be reactive and only come on board when others have made it clear that actually not to do so is illegal or unacceptable". That challenge and how it is being accepted have just got to the point where it can be articulated. We have not got close to formulating a response which will allow it to become part of a programme. But we are digging around the edges. The EDI approach will allow the necessary conversations to take place in full.

EDI can introduce difficult discussions to community based organisations whereby people will be encouraged to take action to change, especially if a project is recognised as being very exclusive and does not involve the wider community. Across rural Northern Ireland there is, however, some evidence of community and voluntary sector alliances which increasingly demonstrate a willingness to talk about difficult issues associated with a divided society. In Newry and Mourne district there is, for example, a Community Relationships Forum which draws participation from the District Council and is facilitated by Mediation Network. One observer of this process has commented as follows:

> The Forum brought in all the people who had something to say about contested parades and marches in Newry town. It is now talking about the flight of the Protestant community. The Forum provides an opportunity to talk about things that had never been spoken about on a shared platform in a way that at the same time is building relationships which might create the environment or conditions whereby diversity can be tackled. The Council, for example, has heard messages about itself which it was not disposed to listen to before, nor indeed was there any mechanism for it to hear them. In this context the Forum is a useful mechanism which can absorb flack. The Forum concept as an informal process has value for other places because people will not address an issue unless there is an awareness to do so, unless there is a debate or discussion started. There has to be that vehicle there to start things going. Then people will say "I have heard things in a different way that I have never heard before – I have changed my mind". The conversations can then move on to talk about the nature and extent of sectarian graffiti, territorialities and contested space, and local prosperity... The goal should be to see if we can get all the groups in a town into the same room to talk about the content of a common, integrated plan for their town and what everyone's role might be in a collaborative, non competitive, complementary way of working.

EDI is, therefore, about trying to get people to think and act in a relational rather than individualistic way. It is about building community with equality at its heart, welcoming and celebrating diversity and people learning to listen to each other in an interdependent way. Leadership by a motivated individual can be crucial to the initiation of that local dialogue and change. The characteristics of such a person currently active in a divided town have been summarised as follows:

> This person works with others within his particular context and his overarching vision for the town. He has already recognised that there has to be an inter relationship with Catholics, Nationalists and Republicans. He wants to find a new pathway that is not about UDA flags, the UVF, Catholics being burnt out of their homes and property being damaged. He wants to get away from the seasonal antagonism of land, particularly the nationalist parades and St Patrick's Day. This person, the members of his group and his affiliated groups will start to develop an understanding about how they view the world, how they view others and how they view themselves.

Leadership, therefore, goes on to embrace a group dynamic built around a sense of dissatisfaction about the local environment, perhaps its segregated character and the strong degree to which it is intensely sectarian and limiting in terms of choice and personal freedoms. One interviewee described the following group scenario in this fashion:

> This group is part of the community but it does not comprise the 'usual suspects' from within the community. It has people, for example, who have a business stake in the town. These are people who are task oriented, who want to tackle problems. They have established, not a neutral venue, but a shared venue in the middle of the town that really breaks the mould. (In this regard, neutrality is about nothing on the walls, it is sanitised, it reinforces our society's urge not to talk about things which are offensive or divisive, not to create offense to anyone, and to talk about only what we have in common. But these people need to talk about the things which are most difficult, contentious and divisive, for only in that way will their town start to improve.) Their shared centre will be a place where differences will be celebrated, where there will be an appreciation that here is a space where we can talk about things, here is place we can all go to. The interesting thing is that, as this project matured, other people did come out of the woodwork and say "What are you doing down there? It is very good. Do you need some help?" People would come up and say "Keep me informed. I want to be part of this". In other words, people were quietly stepping out! This group got support from surprising quarters, which it never expected to happen. But they also got hit as well. The place was attacked and burnt down. But they handled that very well. They visited all the homes of those people the next day and talked them through the issues and in this way have overcome a lot of skeptics and doubters, not least among the politicians. EDI in this context is about cultural change!

A crucial element in the promotion of constructive community dialogue is the intervention of skilled facilitation. In one regional town, for example, a project started with a small group of people meeting regularly to reflect on local issues and deciding to do something together. The group, drawn from different faith communities, agreed to bring in a facilitator to assist with the preparation of a 3 year strategic plan. They subsequently received funding for a full time development officer. This has allowed the exploration of issues with young people such as citizenship, peace making, religious faith and crossing religious divides. They have recently opened a drop-in centre in the centre of the town. The initiative, with its structure, its presence, and its drawing in of people from all religious traditions, is perceived as making a substantive difference compared to other news about the town which is making the headlines in the local newspapers. The involvement of an independent, outside person is regarded, therefore, as important in building confidence to hold the necessary discussions on sectarianism. Continuing to train people with the skills to be effective in this type of arena is crucial for the future. However, the depth of dialogue within any group can vary enormously and as one interviewee illustrated with regard to a facilitation experience elsewhere in rural Northern Ireland:

> Members of the group were saying that: "We get on well, relationships are good here". But I threw a few questions at them and soon discovered that they don't really know each other that well, the meetings were fairly infrequent, fairly superficial, and the underlying issues of prejudice, social injustice and sectarianism were not really being tackled. After four meetings the initiative didn't get off the ground. Some people backed off, there was a lot of hesitancy, there were issues around what can we do together and what can't we do, thorny problems like shared worship and theology.

The failure of clergy in that instance to be comfortable with an emergent agenda of inclusion stands in marked contrast to a large group of women from within the same area who have sought to pursue a collective desire for a meaningful set of experiences built around, for example, the establishment of a local forum, a shared celebration of common religious festivals, and a joint Songs of Praise. The significance of this illustration is that it is the women, after a fairly long period of confidence building, who are now pushing the clergy! Empowerment has enabled institutional rigidities to be challenged from the grassroots in a positive fashion.

For many organisations there is a common appreciation that some of their dialogue with rural communities is conducted under very difficult circumstances. Reference was made to having to engage in a number of instances with paramilitary linked groups. It was reported that it was not always possible to progress these initiatives and as one interviewee explained:

> If it gets to a stage where people feel uncomfortable, do not want to move forward at this point in time, then there has to be disengagement because you cannot compromise your staff and that group cannot compromise itself in its community. But we would say that perhaps we could work with them in the future and we do signpost the way ahead as best we can. This is a difficult area and it is going to get more difficult in the context of new programmes.

However, it was also pointed out that where there are statutory service obligations to be met, any disengagement from a contested local community arena can be especially difficult, notwithstanding perceived vulnerability by staff.

The types of relationship between different rural groups can have a deep influence on the inclusionary nature of local dialogue. Thus, for example, there are occasions when residents on village, small town and urban periphery housing estates are regarded as being apart from the rest of the community thus breeding a sense of hopelessness based on abandonment. These perceptions based, perhaps, on the reality of greater deprivation, benefits dependency or physical environment characteristics have required that service providers come together with local people to draw up an action agenda, including the sharing of basic information. The underlying alienation was well described by a field worker:

> People in that estate say that they have been left to themselves and that nobody cares about them. They say "A lot of us have been employed in the big companies. When they pulled out, we were discarded. We had originally been moved, many of us from rural areas, into these estates to service the industries. But there is nothing there for us now, and nobody seems to be caring what's happening to us. Now we didn't want to be seen taking money off the state. Taking benefits doesn't seem to be the right thing to do. We have been brought up to be loyal British subjects and you don't go and take money off your government. But what are we to do?"

Inter-organisational collaboration

Inter-organisational collaboration is seldom an easy task, not least within Northern Ireland. One interviewee outlined an interesting scenario with implications for EDI:

> There was a sectarian prejudice about how they perceived my organisation. But that organisation with the wrong perception had not gone through the EDI process itself to realise that the organisation that they felt was sectarian had gone through the process! In every organisation you will find some part of the core into which EDI fits but it might take ten years to realise change.

Within different localities there may well be a raft of initiatives which seek to embrace aspects of EDI related work. The issue here is the extent to which there are organisational cross-overs which begin to create a larger scale of activity. At present in the words of one interviewee:

> There is not enough evidence of this around yet, but certainly there are the beginnings. There is a growing awareness in many areas that there are other groups trying to make progress. In some instances they are already in collaborative efforts.

Collaboration at the inter-organisational level was interpreted by intermediary funding bodies as having implications for EDI which stem from the special relationship with a core funder. One interviewee outlined some of the possible challenges ahead:

> We have not engaged in a debate on EDI yet with other organisations with whom we work. When we get a service level agreement from our core funder to administer programmes it will be interesting to see what criteria they lay down, what sort of impact indicators they identify in order to evaluate our performance. If at a practical level they decide that a certain percentage of their programme monies is to be spent in disadvantaged areas rather than on disadvantaged people, that will obviously impact on our EDI policy. Our spend has to be broader than disadvantaged areas, it has to include disadvantaged groups. The constraints that funders impose will impact on the EDI policy of an organisation and while it is important in the current climate to be flexible, I wonder if the funders have themselves have had this debate? Our closest colleagues are in the Rural Community Network and working EDI with them should not be a problem. They are probably further down the line on this, in exploring its significance, than our organisation would be. The level of resistance among District Councils and Partnerships will depend on the political make-up of those groups. But we will have to address EDI at a value level. In other words there are certain values or principles that are core to our organisation's strategy. These core values are about social inclusion, targeting social need, partnership, respect for difference, innovation and valuing people. They are fine in a strategy document but we need to tease these out when it comes to drawing up detailed programme complements. So this is the programme, these are the measures, these are the selection criteria for that measure – in other words this is how our organisation will assess an applicant. Now if we introduce selection criteria that directly or indirectly exclude certain New TSN groups then we will not be in compliance with the relevant principle. The proof of the pudding is going to be in the eating and we will have to ensure that an EDI philosophy has relevance not only for the high level core objectives and values of the organisation, but that it permeates right down to programme measures and selection criteria. This will have implications for our monitoring and evaluation. But this is where the influence of funders also comes in. If they say we want to assess your performance on this particular framework and we are not too concerned about your EDI position, we have to ensure that EDI is accepted as a legitimate performance indicator. Of course our funder is likely to say "Well why would we not want you to do that?" We need to ensure that it is given importance. Our EDI approach has, therefore, to be built around the capturing of the social and economic outputs and impact of the work that we do. We need to think about indicators on the EDI side.

The pivotal position of Rural Community Network, as an "open, friendly and approachable organisation", in driving forward an EDI framework was recognised by a number of interviewees. This recognition is built around the multi-level connections of the organisation: down to rural communities and the Sub Regional Rural Support Networks, across to other intermediary bodies including LEADER Action Groups and District Partnerships, and up to District Councils, agencies and central government departments. Particular mention was made of being able to work more effectively at local level by drawing on the rootedness of the Sub Regional Rural Support Networks, although a call was made for RCN to do more work "on the ground" which is not always connected to public policy matters. EDI training could be one such avenue of activity. RCN has introduced EDI related practices for its own staff. The rationale, future direction and implications of this work were usefully expressed as follows:

> The operationalisation of EDI has been very deliberately about doing the work internally first. There is a degree of risk and thus we are doing it ourselves before we ask anyone else to engage in the process. Internally we started with the senior management team. We did bring in an outside adviser who suggested this approach. The alternative would have been like saying: "We will do community relations work with young people and community groups, but we will not do it with our management committee". From that senior level the work will cascade down through the organisation, out to sub regional networks and their members and then, hopefully, to influence some of the other rural stakeholders such as Rural Development Council, Rural Development Division of DARD, and other agencies. It is a middle-up-down approach...It is a good time for us to embrace EDI because we have been working together as a team for a period of time. We have changed the culture in the past. Each time it gets a little more difficult. With every recruitment diversity widens. Thus the issues revolve around: how did you get here? Different methods are used to talk about this, for example, a small group of 3 people going for a walk in the hills. The process brought out so many types of diversity that we have in our organisation in terms of our backgrounds, but it also provided an opportunity to talk about those personal backgrounds that you would not usually want people to know. The process creates an openness and raises the diversity issues of which there are many. It then moves into considering what we think EDI is, how we feel about the terms. The next stage is applying those in relation to our work. Interdependence during this process becomes a central product because levels of trust and improved working relationships are built up. It has reframed working relationships. But there is a big trust and responsibility element for each other in taking forward the EDI work... Our experience is that it is very, very powerful, very emotional. It creates an openness to hear things about equity diversity and interdependence as opposed to a defensiveness...The defensiveness that arises in organisations following criticism from their constituencies can be put down to the fact that they have not looked at EDI internally themselves, so they are very uncomfortable with hearing criticism about how they are perceived.

Many interviewees did confirm a willingness to collaborate with RCN on EDI issues, for example, to do some shared theological reflection on the themes of community development and adult education. On the other hand quiet concern was expressed about the multiple stakeholder approach adopted increasingly by RCN to legitimise funding applications along with its conveyed perception that non participation by invited stakeholders could diminish application success. A view was also expressed regarding a perceived lack of presence by RCN on certain issues. The interviewee from the Travellers' Movement queried:

> Where do they stand on Travellers' issues? What are they doing to promote the inclusion of Travellers in rural areas? We have never had a direct approach from them. But we would be very willing to work with them.

At the broader level of organisational interdependence, the existence of a perceived operational tension within the rural family between RCN and the Rural Development Council was highlighted. The greater policy development and information dissemination capacity of RCN was, however, hailed as one of its special attributes.

Conclusion

A recurrent theme in this chapter comprises the dynamics shaping intra and inter organisational relationships in the sphere of rural development. However, the stock of social capital available is unevenly distributed and it is not all of the same quality. Clearly if EDI is to be realised as a transformative approach it needs to acknowledge the constraints as much as the potential of local networks and soft infrastructure. The basic question is "What chance is there of interdependence in a socio-cultural system intent on producing and reproducing self-contained stocks of social capital?" Moreover, the logic is thin or at best not proven to suggest that a single identity community can be developed to a stage where it can engage or negotiate with a given out-group from a position of confidence and strength. The very real prospects that mutually exclusive single identities could be deepened and reinforced minimising any real prospect of interaction, let alone interdependence, is a very real one. The history, for example, of the Orange Order and the GAA in rural Northern Ireland demonstrates the way in which the trajectory of social capital building simply does not crossover in any real or meaningful way. Thus, EDI names the dilemmas which shape relationships in Northern Ireland and needs to articulate a respect-based route map that ultimately can connect these diverse and contradictory dynamics. Only in this way can it then be claimed that the EDI process will not legitimise sectarianism.

Furthermore, in what circumstances and in response to what settings one approach provides a way forward as opposed to another, and what the cost implications of this might be, need to be unravelled in the delivery of an EDI initiative. Considerations revolve around perspectives of spatial differentiation and organisational possibility. Clearly rural Northern Ireland is in flux and thus EDI offers a choice to organisations as to how they reshape themselves. It represents an option to embrace inclusion which is removed from the old mode and which goes beyond legislative parameters of conduct. What is crucial is having a number of influential change agents within organisations who can operate close enough to the status quo and with each other, but also bring about change in a non threatening way. In each case the principles should be the same, although how they are worked out in each organisation will be different. The strongly held view is that this will not result in the creation of an EDI culture where everyone is the same, but rather with the situation where everyone's organisational culture has to change as a result of going through a dialogue process which is sustained over the long term.

Chapter 5

Perceptions of Diversity and Inclusion Among Northern Ireland Area-based Service Organisations

Introduction

The previous chapter has presented evidence around issues associated with equity, diversity and interdependence largely from the perspective of Northern Ireland-wide organisations with an interest in rural development. However, the governance of rural development is also fractured spatially at the sub-regional and district scales. Research indicates that there are multiple structures/organisations operating within the same local territory, that these often pursue overlapping objectives and generate a crowded arena for the delivery of local development (Hart and Murray, 2000). However, a significant driver for action at these levels is partnership between the public, private and community/voluntary sectors. Within Northern Ireland this configuration of partnership-working has facilitated the growth of a substantial civil society, arguably the bedrock for transformation in a divided society. Participatory democracy thus sits alongside representative democracy. Given this state of affairs it is appropriate that this chapter should investigate the EDI challenges being faced and responded to by organisations operating within the diverse set of territories making up rural Northern Ireland. Accordingly, this chapter of the book reports the data derived from a suite of semi-structured interviews with key informants from Sub-regional Rural Support Networks, LEADER 2 Local Action Groups, District Partnerships and District Councils (see Glossary). While dealing with a broad range of diversity and inclusion issues, the narrative moves beyond the analysis of the previous chapter by giving attention to the preliminary identification of good practice implementation guidelines.

Sub-regional Rural Support Network perspectives

This section explores equity, diversity and interdependence from the perspectives of the four Sub-regional Rural Support Networks of Oakleaf, Cookstown and Western Shores (CWSN), Tyrone-Armagh-Down-Antrim Network (TADA) and South Down (ROMAL) as identified by Figure 5.1. The analysis explores some consistent themes across the Network territories in relation to social exclusion, the challenges for developing an effective EDI approach in a rural setting, priorities for any future programme and some related implementation considerations.

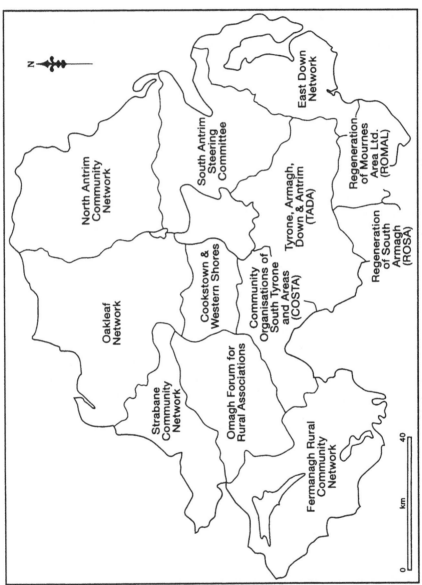

Figure 5.1   Sub-regional Rural Support Networks in Northern Ireland

Despite the differences in the geography, economic profile and community balance in the four areas, there are common challenges to EDI across rural society which transcend religion. These include widening economic disparities reflected, in particular, within the housing market, the problem of close and closed communities, and the legacy of violence in divided rural communities. A particular trend identified across all four Sub-regional Rural Support Networks is that the economic growth that accompanied the peace process has been unevenly distributed with sections of rural society benefiting from new job opportunities and improved wellbeing, but with others falling further into poverty and deprivation. A small but significant group that is benefit dependent, with low employability and few resources such as a house or a car, experiences poor quality of life in some rural areas. In South Down the housing market is subjected to pressure from second homes, proximity to the Belfast – Dublin growth corridor and by its location within the Belfast Travel to Work Area. Affordability, especially for local first time buyers, is becoming a significant problem and has emerged as a priority issue out of the consultation process associated with the preparation of the Banbridge and Newry and Mourne Area Plan 2015. The "equity" dimension to EDI will need, therefore, to address widening social division as well as religious and political diversity in rural areas.

All those Sub-regional Rural Support Network co-ordinators who were interviewed, believe that any EDI approach must move beyond religion to embrace new and more unpredictable forms of exclusion. A consistent theme across Networks was the needs of young people and especially young males in rural areas. Suicides among males is a growing issue, their problems are fairly hidden and support networks are comparatively weak as one interviewee highlighted:

> We have been trying to put on training to improve confidence and address the problems of young people such as drugs and alcohol but they are a difficult group to get at and often it is too late even if you can intervene.

Domestic violence and child abuse are significant problems that are often denied or suppressed. Getting community and organised groups to engage these issues will be a difficult and slow process, but there is a clear need to draw these concerns into the broader design of EDI programmes for rural areas.

The group of Network co-ordinators felt that community relations are improving in their areas primarily as a result of the peace process. Protestant community infrastructure has developed and, despite continuing disparities, there is a perception of social movement, especially with evidence of inter-group contact. Their specific perception is that the Protestant population is not as precarious as it had been in the past within border areas, though others may take a different view. An interesting point made in two interviews is that many of the new housing developments provide a safe environment for Protestants and Catholics especially in areas of high conflict. The implications for EDI are also important. In the context of peace and some form of political stability, integration and mixing might be happening slowly, but it is perceived to be "at least going in the right direction". This has been especially important for the work of the networks with young people and their emphasis on building contact, sharing experiences and collective campaigning. EDI can build on this more optimistic mood.

Nevertheless, one of the imperatives for integration remains the changing nature of the farm economy. The restructuring of agriculture, BSE and Foot and Mouth Disease have produced crises of confidence in the sector. Moreover, reluctance to trade land across the religious divide is still a problem, but economic necessity is driving change as these comments illustrate:

> If a Protestant farm went for sale, Protestants were encouraged to buy it... But that attitude is melting, especially as people have less attachment to land, community and Church.

This restructuring of rural society is also seen in the emergence of stronger interest group politics and local campaign issues related to the rural environment and economy. Moreover, Protestant community groups have seen the practical benefits of local development for other areas and now want to replicate this experience in their own communities. This may mean building on expertise already assembled in Protestant areas and a number of rural regeneration illustrations were offered by interviewees to illustrate this evolving participation in community-led development. Within this context it was suggested that there is some sense of optimism about the ability of rural communities to "naturally" reach out to each other:

> A Protestant in rural Cookstown has more in common with a Catholic from the same area than with a Protestant from East Belfast.

A key obstacle to the implementation of effective EDI strategies is the closeness and closure of rural communities. Certainly the closeness of communities and their sense of identity have helped to build local organisations and social capital, but this can also promote exclusive approaches to development. A clear problem is the spatial segregation of rural areas. For example, in the Cookstown and Western Shores Network territory, the Lough Shore and the Sperrins are predominantly Catholic (say, 75%), whilst the central belt around Cookstown is primarily Protestant. Here the ability to reach out in an "interdependence" sense will be guided by the pace that any one community is prepared to go:

> People are in a personal comfort zone of standing still. Why would they do this work when it has so many risks ... particularly from within their own community?

Moreover, this sense of closure predates the most recent Troubles and formed part of the culture, memory and landscape of rural Ulster. These are, therefore, deeply structural problems that need to be recognised in the design and delivery of EDI in rural areas. In a study by the Northern Ireland Rural Development Council (1997), titled *Sense of Belonging,* attitudes to areas were measured. The results

Table 5.1   Do You Feel Out of Place Living Here Because of Your Religion?

Area	"Yes" (%)
Four Network Areas	9.5
CWSN	14
Lough Shore	4
Stewartstown	19
Moneymore	19
Pomeroy	28

Source: Rural Development Council (1997) *A sense of belonging in Cookstown and Western Shores Network, Fermanagh, East Down and Oakleaf. Cookstown:* Rural Development Council.

demonstrated that, at that particular time, people living in the most contested areas had least attachment to place. As illustrated in Table 5.1, close communities such as the Lough Shore (Figure 3.1) had a low rate of "feeling out of place" compared to more segregated villages such as Pomeroy, Stewartstown and Moneymore.

All of these insights have implications for the way that an EDI programme would be implemented. The challenge for EDI is to articulate the connection between respect for diversity whilst encouraging interdependence between the two religious communities. Working through organisations that are shaped around distinctive religious and cultural values has important benefits, but where, when and how connections will be made to move beyond internal community binding towards a broader bonding of rural society will be crucial. Thus, the Orange Order and the GAA are strong in all the areas reviewed and each represents a significant institutional capacity in Protestant and Catholic communities respectively. The Orange Order has recently taken some initial steps to develop a community strand to its activities. It was also noted by interviewees that Protestant groups are more likely to embrace single identity work, whilst Catholic groups are more tuned into peace and reconciliation initiatives. Thus the issue of whether single identity work simply reinforces difference and limits the possibilities for interdependence needs to be part of the risk assessment in EDI planning. This in turn has implications for traditional interpretations of community relations work which are often treated as a sanitised interaction between Protestants and Catholics.

But there are other dimensions to community seclusion. In some areas, where middle class interests dominate the community and voluntary sectors, very different processes operate than in areas with a more locality centred approach to local development:

> Social class differences are very important between the 'haves' and the 'have-nots'. The 'haves' do your work for you. They are in the driving seat and you are a passenger on their bus – you have to ask permission to get on.

One interviewee pointed out that in those areas "driven by a class based approach", religion and political differences can be difficult to raise in the development process:

> Polite avoidance is still the hardest thing to get around. If the issue isn't raised, then it isn't an issue.

Social class differences emphasise the fractured nature of rural society. Inter-community differences are well documented and acknowledged, but intra-community cleavages also present practical obstacles to the effective implementation of EDI strategies as the following comments illustrate:

> We have had to bring in mediation to integrate the work of churches in Protestant areas. In one case six different churches had groups, were looking for halls and were also trying to get people involved. This was pulling people in different directions and it was very hard to engender community spirit.

> In Nationalist areas there were splits between the middle class and the working class, between the SDLP and Sinn Fein, and between the Provos and INLA. Sometimes it can be hard to find out what is going on, never mind trying to do some work.

A concern voiced by all interviewees from the Sub-regional Rural Support Networks is the manner and pace with which EDI might be implemented. These key concerns included the following:

- people would feel that they could be labelled when they would rather remain colourless;
- it might detract from the core purpose of local activity eg. the needs of young people or farmers, by focussing on political and religious differences;
- natural integration through personal contacts might be compromised if EDI is forced and not allowed to happen at its own pace;
- EDI should underpin all aspects of work and should not be seen as a separate activity for the Networks;
- finding closure to processes that raise political and religious issues within and between groups.

In a similar vein one interviewee pointed out that Network members who were in the Orange Order met in a former Catholic monastery; this may not have happened as easily "if fingers were pointed at it". Moreover, the core business of local development might be compromised or even de-railed by EDI. Thus, another interviewee suggested that if the network attempted "to raise policing issues, half of the group would leave and then blame us for getting them there on false pretences". This sense of needing to tread very carefully was voiced in the following way:

> Yes, we have raised this (community relations) issue with groups...On one occasion it ended up doing more harm than good because we had no way of resolving the issues and problems that groups raised... It put back personal relationships that we are only starting to re-establish now.

In short, it is clear that exit strategies need to be thought through before EDI work is implemented and that very focussed work on outcomes identification, risk assessment and contingency planning is conducted as part of the programme design exercise. The development of skills in conflict mediation and dispute resolution is regarded as a priority by Network personnel. Interviewees wished to see EDI give prominence to generic skills for building on the cross-community connections that are beginning to open up. There is a need to deal with conflict and trauma while at the same time move people on from the violence of the past. Methods to deal effectively with inter and intra group disputes are increasingly required by Network co-ordinators and additional practice assistance and guidance are called for in order to be effective in this difficult arena. One interviewee succinctly expressed this combination of frustration and desire in the following way:

> Some community relations experts keep their knowledge and their skills to themselves. It is a 'black box' and the rest of us are not allowed to see what is inside!

LEADER 2 Local Action Group perspectives

The EU LEADER 2 programme funded the work of 15 Local Action Groups across rural Northern Ireland over the period from 1994 (Figure 5.2). It is intended that the current LEADER+ initiative will form an integral part of the broad range of support from the Department of Agriculture and Rural Development for rural areas out to 2006. It is against this context of involvement and continuity within the rural development arena that research interviews were convened with representatives of four LEADER 2 Local Action Groups. These comprised: South Down and South Armagh, Craigavon Rural Development, Cookstown LEADER Programme, and Roe Valley LEADER Group.

Each LEADER group is managed by a partnership board drawn from the public, private, community and voluntary sectors. Each interviewee underlined the enthusiastic commitment of board members who were seen as comprising a "a broad church" of rural interests and values and who responded to the challenge of identifying and supporting projects. However, it was also suggested that because the emphasis throughout was on spending the budget, participants perhaps had given less weight to developing strong relationships among each other. As one project officer put it:

> I am afraid that the people on the Board really don't know each other, even after a couple of years of sitting together. They read the minutes, they talk about them and the projects, but there are no other interactions. When they come along they do the business, which is really good, but there is a whole other area of relationship building which needs to be explored. The Board should be more than a clearing house for grants.

It is interesting, therefore, that in trying to deepen group relationships mention was made of the importance of a training video, Tips and Traps for Funders,

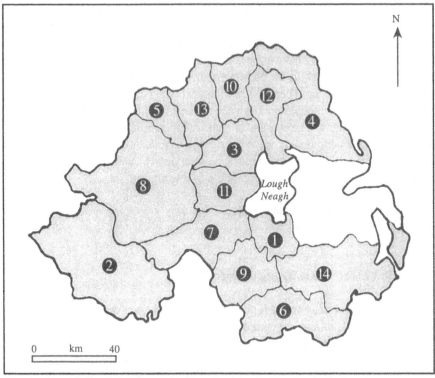

Figure 5.2   LEADER 2 Local Action Groups in Northern Ireland

1	Craigavon Rural Development
2	Fermanagh Local Action Group
3	Magherafelt Area Partnership
4	North Antrim LEADER
5	Rural Area Partnership in Derry
6	South Down/South Armagh Local Action Group
7	South Tyrone Area Partnership
8	West Tyrone 2000
9	Armagh District Local Action Group
10	Coleraine Local Action Group (COLLAGE)
11	Cookstown LEADER
12	Lower Bann Local Action Group
13	Roe Valley LEADER
14	Rural Down Partnership
15	(not shown on map) — Canal Corridor Partnership

prepared by Future Ways. This seeks to introduce some of the conceptual material associated with EDI, but without using those terms. Thus recognition is given to the reality that it can be difficult to get people at Board level to talk openly about these matters and that even if EDI is tabled "there will still be people who will not want to talk about it or address it".

From a beneficiaries perspective it is argued that the EU LEADER programme has allowed a greater number of people in rural society to interact more widely with each other thus beginning a process of challenging isolation and individualism. One Local Action Group decided, for example, to target farmers with a small grants scheme for stock improvement and soil analysis. The outcome over and above the delivery of a set of products was described as follows:

> These small grants often brought the farmers into contact for the first time with the District Council, with the officers of the Council and the Council building, as well as the officers of the LEADER group and the wider LEADER organisation. For a small grant of £250 farmers would lift their sights into the wider rural development sphere. It allowed farmers to interact with other elements of rural society. Farmers are usually very individualistic or they go through the Union. LEADER gave farmers confidence to engage more widely. This spilled over into some involvement with community groups. Now some farmers are saying that they would like to be involved with the new local action group. Thus, while previously being very insular, they are now prepared to take an interest, even beyond the district.

Similarly, the establishment of tourism networks brought people together who would have remained disconnected. The point was emphasised by interviewees that conventional quantitative monitoring and evaluation has missed out on these significant dimensions to LEADER. This observation has implications for EDI with its quest for qualitative changes in relationships. The culture of evaluation will have to be reshaped if only because:

> the bean counters will miss the point of EDI! This cannot be measured on a 1 to 10 scale and it is definitely not about creating, say 10 jobs!

In a follow-on question the contribution of LEADER to reconciliation and integration within the divided society of Northern Ireland was explored. Interviewees identified this as being more within the remit of the EU Programme for Peace and Reconciliation and that LEADER had not overtly embraced that agenda. Probing further into the potential for breaking down sectarianism through rural development generated a number of different perspectives. On the one hand the view was expressed that the need to address shared problems, such as BSE or Foot and Mouth Disease, will marginalise sectarian attitudes and behaviour. On the other hand the following perception of rural life was offered by a LEADER officer:

> We are awfully polite in Northern Ireland. People will say "Oh there is no trouble here, we get on well!" But when we are amongst our own, what I have heard said is not what was said previously in a public meeting because people are not yet safe and

secure even a number of years after the cease-fires. People cannot say what they truly believe. I have had the experience of nasty things being said on a one to one basis, but when it comes to a collective basis, people are very nice to each other, especially when there is an outsider present. In other words, they may say things are rosy, but when you scratch the surface, the old sectarian element is very strong. Rural development has played an important role in trying to lessen that.

While interviewees were enthused by the contribution of EDI to organisational and societal change, considerable caution was advised in bringing forward an EDI programme. One perspective identified the ease with which this could be perceived as another "do-gooder" scheme and that the practice could do "untold harm" if it were to be devised and implemented in isolation. There was strong consensus that EDI cannot be mandatory and that, at best, any funded programme can only provide space and opportunity. Nor can it be left as a "feel good factor" otherwise it will quickly dissipate. One interviewee argued that EDI should not be an add-on to anything else and that even the idea of an EDI training programme is itself incorrect. That comment was illustrated by the perceived way that Government policy makers operate:

"We as policy makers will write our programme and stick on a women's initiative at the end to show that we are consciously targeting women". All policies and proposals should in the first instance be more sensitive to women!

In other words the view is that EDI should not only be worth doing for its own sake, but that it needs to "be at the centre". It certainly needs to be welcomed by Government departments and within the European programmes, such as LEADER+, it could be embedded as a capacity building measure, though Boards might respond by asking: "Do we have to do this? Is it essential for our funding? If not, perhaps we won't bother". However, EDI should not be regarded as yet another rural development training programme. It has to be a learning experience "of good quality and meaningful and has to involve expert facilitators rather than the usual consultants". There is an imperative to "train the trainers" and if it is intended to run this out across community groups, there has to be an appreciation of the reality that "community groups are trained out". This raises the issue of language and some interviewees suggested that equity – diversity – interdependence could "turn people off". On the other hand there was the reflective suggestion that EDI will mean different things to different people and that if the word "sectarianism" can be regarded as vocabulary of the past, equity, diversity and interdependence could provide a new and more comfortable vehicle for people to express themselves by. One interviewee warned against EDI rhetoric and cynicism and suggested the following scenario:

EDI principles fit well with the mission statement of any community organisation which wishes, for example, to build a new hall. To get funding it may say "We will include a statement that we aspire to bring all traditions together". But at the end of the day everyone knows that they are not that bothered. All they want is a nice, shiney, new building. The reality is that community groups are getting cleverer and cleverer; they know the game, they learn the words and the language; if EDI is held up as an aspiration you will quickly see them dancing to that tune also!

In short, there is the opportunity and need to capture the imagination with something that is "different, innovative and personal attention grabbing". This challenge was neatly captured by one interviewee:

> We need to embrace prejudice identification first, before we can move on to prejudice reduction. But regardless of how you do it you will attract the people who are interested or converted. The people from the grass roots do not come to sessions like this. The same 15 to 20 people will turn up for a training programme, but the people living in isolated rural areas or in the housing estates who are dealing with sectarianism on a daily basis are not the people attending. They are the ones you need to be there, and I really don't know how we get to those people.

Rural development should, accordingly, be concerned with reaching out and bringing people together. The fact that LEADER 2 has made a contribution on this front, as noted above, prompted a measure of concern that the Northern Ireland emphasis of LEADER+ on micro-business is a shallow response to the much needed community bridging and bonding work ahead. Local Action Groups have the potential to do more on these fronts.

Finally, the involvement of Rural Community Network (RCN) in this sphere of activity was perceived as a reflection of its wider acceptance in rural society and as being in contrast to the prescriptive intervention of a government organisation. RCN was, however, regarded as having few credentials to go into the private sector environment though the insights from an internal organisational EDI pilot should command wider interest. The view was forcefully expressed that rural development support bodies should become champions for EDI and that only on the back of that commitment and experience should EDI outreach to community groups. In looking further ahead, one interviewee summed up the nature of the required long term commitment:

> The really big question to be faced is that of going beyond the community groups to wider communities! But the more people you involve in the earlier stages, the greater is the potential to reach out. People may open up to EDI if there are enough people working together on it. But it will take time. If the work is started, it has to go beyond 2 or 3 years.

District Partnership perspectives

The EU Special Support Programme for Peace and Reconciliation in Northern Ireland and the Border Counties of Ireland was officially agreed in July 1995 following a widespread consultation process. Its two core objectives were to promote the social inclusion of those at the margins of economic and social life, and

1   Antrim Borough Partnership
2   The Ards Partnership
3   Armagh City & District Partnership
4   The Peace and Reconciliation Partnership, Ballymena Area
5   Ballymoney District Partnership
6   Banbridge District Partnership
7   Belfast European Partnership Board
8   Carrickfergus Together
9   Castlereagh Partnership for Peace and Reconciliation
10   Coleraine Borough Partnership
11   Cookstown District Partnership
12   Craigavon District Partnership
13   The District Partnership for Derry City Council Area
14   Down District Council
15   Fermanagh District Partnership
16   Larne District Partnership
17   Limavady District Partnership
18   Lisburn Peace and Reconciliation Partnership
19   Magherafelt Area Partnership Ltd
20   Moyle District Partnership
21   Newry and Mourne Peace & Reconciliation Partnership
22   Newtownabbey District Partnership
23   North Down District Partnership
24   Omagh District Partnership
25   South Tyrone Area Partnership
26   Strabane District Partnership

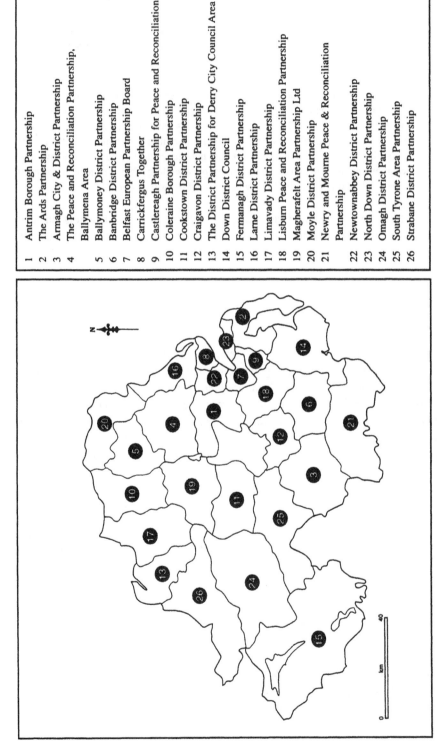

Figure 5.3   District Partnerships in Northern Ireland

to exploit the opportunities arising from the peace process in order to stimulate social and economic regeneration. The establishment of 26 District Partnerships across Northern Ireland (Figure 5.3) was an innovative feature of programme delivery, not least because of their broad membership and responsibility to allocate funding in line with an agreed plan of action (Hughes *et al.,* 1998). The partnership model is set to continue under the EU Peace 2 Programme, although the organisational configuration will see the establishment of Local Strategy Partnerships in each District Council area with a broader remit than District Partnerships. Accordingly, it is appropriate that this research should include reference to the experience of District Partnerships as a vehicle for change within rural Northern Ireland. Interviews were convened with officers from Armagh, Banbridge, Derry and Magherafelt District Partnerships (Figure 5.3).

Interviewees identified a raft of projects which had been supported by District Partnerships to promote social inclusion. Evaluation evidence points to a very high level of inclusion of people in the design and management of these projects. These spanned initiatives targeted, for example, at rural community groups, young people, the elderly, victims of violence, women, people with disability, and farmers. The stories have provided a rich learning experience for the partnerships in regard to what works well and what could be done differently, and in turn have provided valuable insights for this research. Nevertheless, there was an admitted tension between promoting reconciliation and supporting broad social inclusion which was characterised by a genuine struggle to define the former in day to day working practices. In part, this dilemma was viewed as a reflection of language imprecision which interprets reconciliation as forgiveness and thus brings church-based concepts into the socio-political arena. It was suggested that EDI could represent a new dialogue to more effectively get at the underlying emotions of loss and resentment. Only on a few occasions did the intensity and sensitivity of the work of the partnerships spill over into organisational relationships and decision-making. One interviewee described the following experience and identified the implications for EDI:

> We did have a lot of differences, the whole thing blew up and it was very heavy. Much of it was around sectarianism and it was nasty. Some people left the partnership and we tried to handle it, but it left a hiatus. In truth, we were all emotionally exhausted and scared by it, and we didn't do anything about it. We didn't sit down and talk about what the problems were – we just motored on. The whole thing was so sensitive. Well we at last decided to leave it as part of our past and to move forward. Any discussions about EDI in that environment would have been very difficult since people would not have been willing to expose themselves.

In short, the point was made on a number of occasions that people sitting on partnerships do not perceive themselves as buying into a personal and organisational critique. They are there to deliver projects.

Interviewees expressed considerable dissatisfaction with those projects which came forward on the basis of "cross-community fixing" whereby token Catholics or

Protestants were included in the funding bids. There is a strong belief in the integrity of single identity work and the view was expressed that in the future there should be a closer relationship by the EU Peace Programme, for example, with the Orange Order and the GAA. This should not necessarily involve capital support, but rather working with the members of each organisation to help them to more fully appreciate their own culture and that of others. The opportunity to build in better awareness to project support was raised by one project officer whose Board was reluctant to follow that route:

> We gave some funding to a perceived Protestant organisation for its hall and we talked about putting a condition on the grant that the members would draw up a community relations programme. In other words there should be a requirement to learn a little more about themselves and their organisation. Then, perhaps, they would do something to learn about their neighbours of the opposite religion. Well, the District Partnership board would not agree to these strings being attached. But the funding was given!

The importance of training for District Partnership Board members was underlined during the interviews though it was accepted that there is a tension between making it compulsory and avoiding burn-out through work overload on volunteers. One partnership, for example, ran a training course for its Board on prejudice awareness but only half the members turned up. The comment was made:

> It is a shame that the people who need this do not turn up and that they are not comfortable in accessing the training.

Interviewees supported the need for training within the new Local Strategy Partnerships and suggested that if EU Peace funds are going to be administered, then there should be an understanding by people of what the challenges are all about. There should be something included on the themes of reconciliation and community relations. Only then, for example, can the administration of the EU Peace monies make the claim that there is a big difference between this work and that of, say, the Lottery funds. Training, therefore, has to be relevant and should seek to improve effectiveness; it cannot be diluted to being a moral aspiration.

Considerable interest was expressed in the potential of EDI to assist with the organisational development of Local Strategy Partnerships and constituent working groups. At the formation of the District Partnerships the strategic planning process was used as a mechanism to build team capacity and leadership. Now that this work is well advanced through the efforts of programme officers, there is interest that EDI could be an appropriate partnership process. The suggestion was made that the European Programmes Office could take a strong lead on turning this possibility into recommended practice by allocating funding for dedicated partnership capacity building related to EDI, though the title may need to be "dressed up a little". The relevance of EDI, it was pointed out, has never been greater since the focus of the new Local Strategy Partnerships is on integrated planning and delivery across all statutory bodies at the District Council scale.

Within EDI it is the theme of interdependence which particularly resonated with interviewees. District Partnerships in moving to a programme-led delivery approach for the second tranche of spending over the period from 1998 have been required to work through service agencies. The timescales were perceived as tight and in one instance the reluctance of people to come forward to participate was deemed to be "a scary time". Having appointed a contractor to deliver an initiative in a largely Protestant area, it took a further 12 months to get people to sign up. The delay on reflection can be put down to poor community infrastructure, unfamiliarity with the community development approach and a reluctance to ask for help conditioned by a culture of self-sufficiency. The ability to interact effectively with service providers and with key voluntary organisations emerges as a great strength of the District Partnership approach and at the very least the voluntary groups provided a useful "sounding board to bounce ideas off". On the other hand reference was also made to the issue of turf protection, perhaps more symptomatic of a perceived 'crowded playing field' condition at local level. The following comments capture that interdependence frustration:

> Our disability programme just did not work out because of too many gatekeepers. I could never get talking to the people that I needed to because they were always being protected. The disability awareness programme was effectively blocked by the attitude "Don't spoil our work". The response from the voluntary sector and statutory agencies was always "No thank you – we do our own thing on that front. We don't want to know you!"

> We were trying to get into some of the youth clubs but in one instance we had to get past the gatekeeper who said, "No, the kids here would not be interested". But the kids under a different scenario would indicate their interest and when asked their opinion if they were to be approached, they said "We would love to do this!" Through sheer persistence we worked with the schools, then we got into the youth clubs, and now a programme is going really well.

It was argued that EDI requires intensive person-to-person dialogue and that it is very time consuming and labour intensive. In reaching out to interest groups, such as young people, the perception is that this will require unconventional approaches in order to avoid fear and suspicion. The sectarian undercurrents may require the equivalent of "standing on street corners and waiting for them". In other words the necessary confidence building has to be done on their terms and in their environment. The following comments illustrate the scale of the perception gap to be bridged:

> People will say, "We have no problems here". Yet there could be a group of young lads beating the seven shades out of each other next door because of their religion. But they will not recognise that sectarianism is there.

The challenge is to break into this context in a subtle way and to get people to face it. Moreover, it is not about "educating the working classes". In building active citizenship, it should permeate all levels of society.

District Council perspectives

District Councils have increasingly taken a more prominent role in local development, as well as promoting equality and addressing social exclusion. This section is set within that context and reflects some of the experiences of District Councils with EDI, especially with regard to the interpretation of local problems, the content of strategies and programmes, the structural form of local government, and the changing skills required by officials. The research deals with the experience of Coleraine Borough Council and Newry and Mourne District Council each of which, in association with Counteract and Future Ways, has advanced an EDI framework within its organisational setting. In addition, the section includes insights offered by interviewees from Craigavon Borough Council and Limavady Borough Council (Figure 5.4).

One of the constant themes to emerge from research at locality level is the socio-cultural fabric of rural life and the challenges that this poses to the implementation of EDI. This is particularly the case when traditional values and routines have an impact on equality considerations as with the following illustration from a local government official:

> Farms have been handed down to the eldest son for generations.... Women lose out. Other children lose, and families and even whole communities can be torn apart by this.

The meaning attached to land in Northern Ireland is deeply symbolic in terms of family, gender and religion. The complex interplay of social forces, accepted informal rules and kinship behaviour produces a system of succession which can have inequities and unfair outcomes. There is, therefore, an issue as to how EDI can respond in a rural setting to that type of baseline condition.

Within the organisational setting of a District Council there is a set of equally influential social forces, informal rules and personal relationships which have implications for EDI. In dealing with these realities an incremental methodology has been advanced within one local authority, in which progress is dependent on the implementation outcomes achieved in a series of discrete phases. Some of the key elements include:

- the establishment of an EDI Working Party at a senior organisational level in order to develop a specific EDI strategy and to map out its implementation;
- the attracting of political support on an individual party basis. Here the concepts and outcomes need to be negotiated with rather than simply explained to the different political parties;
- approval of the approach by the Senior Management Team. This sequencing did cause some tension, especially as politicians were approached first. But the EDI Working Group took the view that the Senior Management Team might not have approved the progression of the approach on the basis of an assumption that politicians would not accept it or buy into the related principles;

Figure 5.4 District Councils in Northern Ireland

- the bringing of the EDI approach to the full Council for consideration and formal adoption;
- the conducting of a scoping study to identify strengths and weaknesses and to prioritise activity areas for the programme;
- the completion of intensive staff briefings, training and the formation of key sub groups, especially within the trade unions and middle management;
- the establishment of an EDI Development Group to prepare a draft plan of work;
- agreement on the plan and its implementation along with built in monitoring and review procedures.

The plan of action which has emerged out of this process is concerned mainly with the internal environment and in broad terms it looks at the following aspects of the District Council's work:

Factor:	**Content:**
Wider environment	External environment
	Government policies
	Political context
Political leadership	Party political issues and perspectives
	Common purpose role / defining areas of
	co-operation and trust
	Guardians of the process
Council mission/values	Civic leadership
	Good relations
Organisational culture	Tacit culture
	Values and assumptions
Community Forum	Consultation
	Feedback
	Partnership

Each functional department set out a plan of action against each of these criteria. Of course not all these issues are relevant to all departments, but there was an emphasis on devising a consistent framework that allowed the organisation as a whole to produce an integrated, corporate approach to EDI. In addition, this approach made it easier to evaluate the extent to which commitments were being met and to identify problems and obstacles associated with plan implementation.

Both Coleraine Borough Council and Newry and Mourne District Council are extending EDI outwith their immediate organisational settings. Some key elements in that outreach include:

- establishing community forums or open search conferences;
- engaging in new and more participative forms of consultation;
- supporting new experimental models of understanding and accommodation within the wider community;
- targeting excluded interests, thus providing a link with the New TSN and Equality commitments of the Council.

The Northern Ireland Act 1998 has placed new equality duties on District Councils, while New TSN has had the effect of requiring that resources and policies are targeted. Newry and Mourne District Council through its EDI work has been able to run ahead of the formal equality legislation. In 1997 it established a high level, cross-departmental group titled *Relationships in Equality, Diversity and Interdependence* (REDI), the benefits of which were identified by one interviewee as including:

- drawing attention to the precarious position of the Protestant population and its "exclusion" from large parts of Newry town;
- identifying through critical evaluation some of the bias built into some Council routines that were dominated by a largely male culture, especially at senior management level;
- preparing the ground for statutory obligations imposed under the equality legislation;
- orienting people at all levels of the administration towards values of equality and respect for difference;
- linking different aspects of Council functions more effectively, for example, community services, community relations and equality.

The embracing of the equality agenda by District Councils is perceived as having had an impact on making EDI principles easier to adopt not least through the establishment of new structures. In most District Councils new Equality Units have been introduced and these are regarded by interviewees as having made a contribution to the way that policies are produced and consumed. On the policy production side, for example, Newry and Mourne District Council built in the REDI experience to facilitate wide ranging staff research on how policies are formulated and proofed against equality criteria. The emphasis on Equality Impact Assessments, supported by recent guidance from the Equality Commission, is regarded as capable of adding strength to this broad approach. On the policy consumption side, Best Value has necessitated greater consultation with rate payers thus ensuring that Council policies meet citizen expectations, are fair and are clearly understood. Again with reference to Newry and Mourne District Council, equality research has been closely linked with Best Value preparation and both units are within the same Department. The Council held some 12 consultation forums using the "open space" methodology in order to allow people the opportunity to prioritise the issues most relevant to them. Sessions have been held in different areas on

---

EDI Principles for Dignity at Work

The elected representatives, staff, management and trade unions within Newry and Mourne District Council fully commit themselves to the principles of equity, respect for differences and relationship-building across sectarian divisions.

We accept that everyone has the right to work and live free from any form of intimidation due to religious, political, cultural, or national differences and commit ourselves to ensuring the freedom of all those who work for the Council from any form of discrimination by word or actions.

Representation and promotion of our own cultural, political and religious identities should be achieved in a manner that shows respect for each other, promotes diversity and can lead to creating mutual respect and understanding.

Any attempt to prevent the employment, continued employment or career development of any individual within the Council due to religious, political, cultural or national differences will be vigorously opposed. Anyone involved in such activity will be subject to disciplinary procedures.

All staff are committed to ensuring that their behaviour can in no way make any other staff member feel uncomfortable or victimised because of their religious, political, cultural, or national identity.

Councillors will endeavour to use language and conduct themselves in a manner that makes no other Councillor, the community or members of staff feel belittled or degraded. They will endeavour to engage in respectful politics and avoid behaviour that could cause greater divisions within the wider Council area.

The Council will endeavour to ensure all Council premises shall be environments where anything that identifies a particular community allegiance, that could give offence or cause discomfort to individuals, groups, or the community, would have top acceptable to both 'majority' and 'minority' communities.

The Council will proof the delivery of all services and fund raising against agreed Community Relations principles.

The REDI Development Group will regularly monitor and evaluate the effectiveness of this Declaration and all associated structures, procedures and training. It will engage with elected representatives, management, trade unions and staff on any changes which may be required in the future to ensure continuous improvement.

gender, disability and young people issues with additional follow-on engagement planned with Travellers. Through this process the Council has been able to identify the vulnerable groups in the District as comprising Protestants, women and Travellers. Moreover, this approach is perceived as making the point that a range of interests and groups experience multi-level exclusion from local life, especially single identity groups.

This research, accordingly, has identified that some District Councils are putting systems in place that can facilitate the introduction of EDI. Newry and Mourne District Council has introduced a *Dignity at Work Policy* that sets out a code of conduct to promote EDI principles in a work setting. The declaration is endorsed by elected members, trade unions, management and staff within the Council and because of its significance is cited in the box on the previous page.

The important point is that this Declaration of Principles was implicitly shaped by a desire to better manage the political space and by a recognition that regulation and control would not have achieved the goal being set.

The operation of District Council functions in this manner, together with the equality impact proofing of external policies on the groups defined under Section 75 of the Northern Ireland Act, is perceived as having an important bearing on conventional community relations policies. This is evidenced, as one interviewee put it, by a shift away from "festivals and tea parties to work with single identity groups, solving sectarian disputes and making linkages between groups". In Craigavon, for example, a Public Voices Conference in 1999 highlighted some of the main priorities to be addressed within the Borough. These related to weak community infrastructure in Protestant areas and the integrative potential of youth-based development work. The local Youth Forum has provided a mechanism to facilitate cross-community work, share ideas and develop projects for young people in the area. But advocates of EDI see the approach translated outside community relations to all spheres of Council work. It was stated during the research that EDI should be assimilated into the organisational culture and method of working. In short, it should be identifiable across the range of services at all levels, among staff and councillors and in all functions of the organisation. As expressed by one official:

EDI should be like a stick of rock... Every time you take a bite the letters are there.

At this stage in this section of the chapter it is appropriate to consider some of the attendant requirements and implications of EDI within a local authority setting. An important pre-requisite for the operationalisation of EDI is effective leadership at political and civic levels:

Local councillors have shown a keen interest in developing training in the whole area of civic leadership within local development initiatives. EDI could have a large role in setting out necessary principles for deepening this work in our area.

This comment underscores the role of EDI in empowering key agents of change whether they are community workers, administrators or politicians. The location of these key agents, their necessary competencies and the way in which EDI principles can be absorbed within their environments are significant challenges. The committed involvement of local level politicians is clearly necessary in this interface between political and civic activity.

A second consideration, of importance to community development and community relations activists, is the expressed possibility that EDI might frighten people with more new concepts, especially at a time when they are already overburdened with the implications of the equality agenda. The challenge was expressed in this manner:

> We have got to get EDI into everyday words. People are fairly cynical about equality work and we need ways of bringing this to groups in a format that will not put them off.

The goal must be to "get the ideals of EDI into hearts and minds" and the everyday routines of decision makers and community workers. The manner by which individuals will be "hooked into applying EDI" is regarded as an important consideration in programme planning within a rural context. Suggestions include developing areas of activity, for example for young people, that offer the potential to construct active programmes around equity, diversity and interdependence.

Some of the earlier interviews with key service organisations revealed a measure of confusion about the link between EDI and the equality agenda. From the current research there is evidence to indicate that an EDI philosophy has played a specific role within the District Council setting by helping with the formulation of EDI sensitive policies, implementing programmes and establishing structures that cut across Departments, and evaluating the inputs to policies against EDI ideals thus forging the connection between EDI and Equality Impact Assessments. But there is also a clear perception within Councils that Equality, Best Value and Targeting Social Need are collectively having an effect on the way that policies are formulated and implemented. Accordingly, the point is made that there needs to be clear territorial mapping of EDI, along with where and how it relates to District Council initiatives. The configuration of these linkages will of course vary across local authorities. Thus, for example, one interviewee provided the following illustration:

> Gay men are very excluded in this area and yet they are a target group in our Section 75 plan. But there is no way that you will get this addressed by the Councillors.

The intensity of sectarianism within any area and the political polarisation of Council representation will certainly shape the content of local EDI approaches. In some areas where community relations have historically been weak, there is the view that the whole idea of diversity and inclusion will command less institutional support. In those District Councils "the boxes are ticked and the forms are filled in" but there is little imagination, purpose or resources to support these broad objectives. Some of the problems with community relations work within Councils, it was suggested, can be traced back to its initial development under the District

Council programme. It is within this context that some Community Relations Officers perceived their status as weak and too far down the organisational hierarchy. The fact that the then Central Community Relations Unit left so much scope for District Councils to downgrade this function exacerbated the situation. The view is that this history is significant since EDI structures need to be "almost invented" by those Councils that wish to advance these principles within their own organisations. From the Newry and Mourne District Council experience the clear conclusion is that EDI will be easier to implement where there is high level executive and political support. A before and after assessment of the application of EDI approaches reveals the way in which these have had an impact specifically on community relations work. The Newry and Mourne District Council and Coleraine Borough Council experiences are summarised in Table 5.2.

Table 5.2   An Assessment of EDI

Before the implementation of EDI:	**After the introduction of EDI:**
Support existing work and projects	A more focused approach to planning and intervention
Don't upset the apple cart	Target specific groups and contested issues
Avoid focusing on division	More work on interface issues
Cross community approach in every case	Single identity work and progression valued in the long term planning of projects
Work for groups	Greater participation and involvement of people
Community reconciliation is the output	Developing skills to deal with a more complex approach to dispute resolution and conflict is the output

Nevertheless the two District Councils that have introduced EDI approaches were very keen to identify the real obstacles to change. These are institutional, cultural and personal and, as a reality check, they constitute a prominent element in any analysis of baseline conditions. The obstacles have been summarised as follows:

*   apathy within the community;
*   denial of the problem;

- suppression of discordant voices in the area;
- social wellbeing and wealth being given more importance than good community relations;
- passive support for the status quo, which has gained a momentum of its own;
- an approach to development which still emphasises outputs in 'bricks and mortar' rather than placing value on the processes underpinning EDI;
- poor perception of some aspects of community relations work and its comparatively low status within District Councils;
- the perception that community relations tends to favour a Nationalist perspective and the consequent reliance on single identity methods by some Protestant communities.

Conclusion

This review of area-based organisational perceptions of social exclusion, social prejudice and EDI, together with the analysis of approaches to the formulation and implementation of new ways of doing business, have profound implications for other rural policy environments and organisational settings. The interviews capture an appreciation of the need for change and a broad willingness to, at the very least, take some tentative steps towards embracing a change dynamic. It is noteworthy that some District Councils are looking to EDI processes as an innovative way to transform relationships. Their experience will have a deep bearing on any subsequent programmes brought forward. The next chapter looks more closely at the specific progress on EDI being made by a key stakeholder organisation within the Northern Ireland rural development arena.

Chapter 6

Responding to the Challenges of Diversity and Inclusion – Northern Ireland Rural Community Network

Introduction

Northern Ireland Rural Community Network (RCN) is a voluntary organisation established by community groups from rural areas in 1991 to articulate the voice of rural people on issues relating to poverty, disadvantage and equality. It is a key player within the rural development policy community of Northern Ireland (Murray and Greer, 1999) and is core funded by the Department of Agriculture and Rural Development, with the remainder of its income coming from charitable trusts, membership fees and project income. RCN has over 500 members comprising community groups, Northern Ireland-wide voluntary organisations, local authorities, external bodies, consultants and individuals. RCN is currently working to a number of strategic aims:

(1) Strategic Aim 1: To articulate the voice of rural communities; it has published, for example, a gender proofing analysis of the EU Common Agriculture Policy (CAP) reform, a series of rural development policy discussion papers, a public policy consultation practice manual, and regularly hosts conferences and rural development training.

(2) Strategic Aim 2: To promote community development and networking in rural communities; key elements here are helping to further develop the 11 Sub-regional Rural Support Networks (RSN), capacity building in areas of low community infrastructure, and cross border networking.

(3) Strategic Aim 3: To work towards social inclusion and peace building in rural communities by reducing polarisation, sectarianism and social exclusion; this includes the development of mediation skills in rural communities (over 150 volunteers have enrolled) and a community strengthening small project fund. The RSNs have been a key element in the operationalisation of this strategic aim.

(4) Strategic Aim 4: To support the building of sustainable rural communities; under this aim some 29 building refurbishment or extension projects have been funded by the 21st Century Halls programme and a village hall advisory service is run.

RCN sees itself as a learning organisation with continuous review of its work, training and staff development, and a participative management style. In short, RCN takes the view:

- that rural development must be actioned at a variety of spatial scales;
- that voluntary participation through a community development process can contribute to better planning and delivery;
- that inequality and sectarian divisions in rural areas can be reduced by networking infrastructure and community development;
- that capacity building is essential in building community confidence;
- that openness, accountability and transparency are essential building blocks for the equitable development of rural areas.

Accordingly, a commitment by RCN to equity, diversity and interdependence fits well with the ethos of the organisation and its broader contribution to rural development in Northern Ireland. It is against this context that RCN became a participant in the *Gaining from Difference* initiative led by Counteract and the Future Ways Programme in order to better take forward its own organisational approach to EDI. The remainder of this section outlines this involvement thus far.

Equity, diversity and interdependence – the RCN experience

RCN, as outlined above, has had a longstanding commitment to EDI as expressed through its aims and values related to social exclusion. Its activities over the past 12 years are perceived to have made a positive contribution to change in rural areas with its emphasis on networking and involvement in the delivery of the EUSSPPR. There is awareness of the fact that mostly Catholic community groups have been participants within the rural development arena. In turn, this is believed to have fuelled a perception that RCN is a Catholic and male organisation whose interests are located 'West of the (River) Bann' – a slogan which signifies the separation axis of disadvantaged rural Northern Ireland from a more prosperous Belfast city region to the east.. More detailed information on this and other corporate identity perceptions has been obtained from a questionnaire survey* of RCN members and non members, the results of which can be summarised as follows:

- *perceived religious background:* some 40 per cent of member respondents perceive RCN to be equally represented as between Catholics and Protestants. However, 32 per cent expressed its perceived religious background as "generally Catholic", while 0 per cent suggested "generally Protestant".

---

* A total of 358 questionnaires were sent to members, stratified on the basis of county affiliations. A similar number was sent to non members. A 42 per cent response was achieved from members, and a 15 per cent response from non-members.

Among non members, 22 per cent of respondents stated "equally represented", with 37 per cent perceiving a "generally Catholic" background and 9 per cent perceiving a "generally Protestant" background;

- *perceived political opinion:* some 33 per cent of member respondents perceive RCN as neither Nationalist nor Unionist in terms of political opinion, while a total of 31 per cent perceive equal representation. While 11 per cent indicated a perception that it is "generally Nationalist/Republican" there were, in contrast, no member respondents who expressed a "generally Unionist/Loyalist" perception. This broad structure is similarly reflected by non member respondents;

- *perceived gender make-up*: among members some 14 per cent perceive the organisation to be "generally male", with 13 per cent seeing it as "generally female". Over 40 per cent perceive RCN to be equally represented in terms of gender, though just over 30 per cent are in the "don't know" category. In contrast, 22 per cent of non member respondents perceive RCN as "generally female", 6 per cent as "generally male", and 28 per cent as "equally represented";

- *perceived social class make-up*: some 33 per cent of member respondents perceive RCN as "generally middle class", with 10 per cent taking the view that it is "generally working class" and 23 per cent stating that it is "equally represented". Specific class perception among non members is broadly similar, though only 9 per cent of respondents perceive RCN to be equally represented in terms of social class, while a large 45 per cent indicated that they have no knowledge upon which to form a perception;

- *perceived geographical spread*: notwithstanding the perception by 44 per cent of member respondents that the geographical spread of RCN is balanced across Northern Ireland, a significant 29 per cent take the view that it is "mainly West of the River Bann" and a further 9 per cent that it is "mainly East of the River Bann". Among non members, in contrast, only 28 per cent perceive the organisation as balanced across Northern Ireland, with those seeing its activities as mainly West and East of the Bann being 30 per cent and 9 per cent respectively.

- *perceived quality of the RCN voice for groupings in rural society:* among members the perception is that the organisation does its voice work "well" and "very well" (collapsing these two values where positive scores are dominant) in relation to women (63 per cent), farm families (54 per cent), Catholics (45 per cent), men (39 per cent), older people (38 per cent), Protestants (35 per cent), the disabled (33 per cent), youth (30 per cent), Nationalists (27 per cent), people with dependents (23 per cent), and Unionists

(18 per cent). In contrast, RCN is perceived to provide a "poor" and "very poor" voice (collapsing these two values where negative scores are dominant) in relation to children (30 per cent), Travellers (23 per cent), Gays/Lesbians (15 per cent), ethnic minorities (20 per cent) and people without dependents (13 per cent). There are no dramatic differences in perception among non members.

• *perceived quality of the RCN support to groupings in rural society*: among members the perception is that the organisation does its support work "well" and "very well" (collapsing these two values where positive scores are dominant) in relation to women (58 per cent), farm families (46 per cent), Catholics (42 per cent), men (39 per cent), older people (35 per cent), Protestants (33 per cent), the disabled (31 per cent), people with dependents (25 per cent), youth (24 per cent), Unionists (18 per cent), people without dependents (16 per cent) and Nationalists (15 per cent). On the other hand, RCN is perceived to provide "poor" and "very poor" support (collapsing these two values where negative scores are dominant) to children (26 per cent), Travellers (18 per cent), ethnic minorities (16 per cent), and Gays/Lesbians (14 per cent). Among non members the most significant differences in perception are related to the support given to farm families and Protestants. For the combined values of "well" and "very well" the perceived quality of support given by RCN declines to 24 per cent and 19 per cent respectively.

• *open ended perception comments*: a total of 34 additional comments from members were appended to the questionnaires which are both supportive and critical of RCN. Thus respondents have commented on "the excellent service to one and all", the perception of RCN as an organisation which "supports the farming community and lobbies effectively on its behalf", and the fairness of RCN "in dealing with both sides of the political divide". On the other hand there is a feeling that people do not know enough about the organisation, a perception that RCN is a "closed shop" in regard to Board selection and issue representation, and in one instance, the perception that RCN gives stronger support to prisoners and their families compared to the victims of violence and their families. Several respondents commented on the questionnaire itself with views which range from anger about its content, to lack of experience about the issues raised. Issues running through many of the 15 additional comments from non members are the lack of awareness of RCN and the need for better publicity about its activities.

The questionnaire results do confirm some of the self-analysis made by RCN staff. Thus, for example, steps have been taken to improve the gender and religious balance of staff and board members. Nevertheless, RCN has encountered a degree of internal resistance to 'up-front' sectarianism training. Some individuals would admit, for example, that there were bad experiences on this matter in the past and that they would prefer not to revisit this memory. This openness of opinion is viewed as significant since any effort to fully embrace EDI, it is appreciated, will require considerable courage. The initial steps taken in early 2001 by RCN to introduce an EDI framework were not, therefore, about selling a soft public relations image.

Several parameters were designed at an early stage to guide the direction of the EDI process within RCN:

- an integrated approach which would champion fairness, not only in relation to religion, but also other Section 75 groups, as well as wider diversity issues;
- a desire to go beyond the provisions of the equality regulations which would demonstrate the EDI process as a real commitment to change rather than a commitment to the legislation;
- a willingness to uncover the organisational culture and the subtle messages that resound around teamwork, the roles of men and women, what is talked about, who has power and influence, and tinges of 'orange and green';
- an organisational learning approach where things are not hidden but changed, and where the emphasis is on learning from experience, rather than on blame and punishment;
- support for organisational change throughout the EDI process comprising actionable items such as policies, structures and targets;
- recognition that an internal process was needed before the approach could reach out to the wider membership of RCN.

The internal process commenced with the participation of the senior management team in a number of day-long workshops to discover 'where individuals were coming from' in terms of their personal backgrounds. Issues were explored in relation to how EDI should be defined from a senior management perspective and how it could impact on RCN and other rural interest organisations. A similar format followed on to engage development workers. During their three days of training, thus far, the view is that much has been achieved in terms of strengthening teamwork by sharing experiences, and airing perceptions and grievances not raised before. The involvement of administrative staff, it was recognised, required a different approach to bring them together and, when asked how this might be done, a boat trip from Enniskillen with a different facilitator was agreed. It is accepted that a similar method of engagement and pace of progression for each group of staff is inappropriate, though eventually it is intended that all staff should convene together. Furthermore, while reflections on the process as a whole have been recorded, the details of the discussions remain confidential. In short, the experience is that participation within an EDI framework is organic in character and that a unique momentum is developing around the interplay between confidence and uncertainty about the future direction.

Conclusion

At this stage it is possible to identify a number of good practice insights:

- different approaches to engagement and dialogue are needed for different staff groupings;
- time and space are required away from the office for this dialogue;
- everyone should be introduced at the beginning of the process to the EDI framework on the basis of fairness and inclusion;
- it might have been preferable to have established an internal development group of senior management and other staff at the outset, rather than later on during the process;
- skilled external facilitation is very important to provide space and confidence for dialogue;
- the process has assisted team building and greater openness;
- organisational issues, which may have continued to lie dormant, have been surfaced.

In taking this initiative further forward RCN proposes to continue the EDI process among staff but to extend it to include members of the Board. The results of the perception questionnaire, set out above, will be used to stimulate reflection and discussion and have since been published in the RCN quarterly newsletter. Ultimately, it is hoped that this approach to organisational change can be shared more widely, to include the Sub-regional Rural Support Networks and RCN members. The publication of a workbook on EDI (Rural Community Network, 2003) for use by rural community groups is a significant milestone in commencing that process of dialogue. Finally, there is an appreciation that this must be an ongoing initiative, without closure, if it is to be successful. The RCN experience thus underlines the "learning organisation" character of a dynamic, self-critical and dialogue based EDI process. The next chapter seeks to draw together some key insights from this case study and the wider survey of organisation perceptions discussed earlier.

Chapter 7

A Situation Analysis of Equity, Diversity and Interdependence

Introduction

The previous chapters contain a very detailed account of organisational perceptions, attitudes and behaviours related to the themes of equity, diversity and interdependence. The interview data capture the deep-seated problems facing organisations in their work with rural society. Much of the dialogue is constructed around dilemmas and reflects the search by people within their organisational settings to work out the 'right' answer. But the data also point to what is being achieved to create better and different sets of relationships, internally and externally. The narratives thus represent a collective baseline of perceived challenges and grounded hopes. This chapter identifies a suite of important insights which are embedded in the data and which constitute a synoptic situation analysis of where equity, diversity and interdependence are currently positioned in rural Northern Ireland. It is from this understanding that, at a broader scale, the way forward can be shaped.

Key insights

1.   Rural society is extremely differentiated in terms of perceived inclusion and exclusion by service organisations. This segmentation has sectoral, social, and spatial dimensions. In other words, rural society is not a monolithic grouping. Organisations do perceive themselves as wanting to make a difference in terms of better quality of life to a multiplicity of service constituencies through a combination of pilot projects and training programmes.

2.   The groupings comprising rural society need also to be disaggregated on the basis of diversity and inclusion. In other words, there are multiple and overlapping identities and conditions which impact on groups and on citizens within each grouping. Thus, for example, the experience of women in rural life can be stratified on the basis of farming or non farming background, age, race, sexual orientation, family/marital status, disability, religion, political opinion and so on. The inequalities, which express themselves as varying forms of prejudice, are multi-dimensional and may be related, for example, to isolation, non-recognition, poverty, and powerlessness.

3. The distinction between 'single identity' and 'political identity' is a recurring narrative in the research, especially in the context of highly segregated rural communities. There does seem to be a thin line between the two, which presents conceptual challenges to EDI, especially in determining the value of diversity, whether this is compatible with interdependence, and how they relate to each other in a single framework. There has clearly not been the scope within this research project to unpack these dilemmas. Yet they do require further exploration, not least because of the implications for programme funding and delivery. The fact is that political and religious identity divisions affect the political core of Northern Ireland society in ways that issues of gender, disability, sexual orientation etc do not. The former are fundamental features of the constitutional debate on Northern Ireland's right to exist.

4. Social prejudice, in general, and sectarianism, in particular, are acknowledged as being difficult issues to talk about because of sensitivity and as being difficult issues to quantify because of silence. The more local and the more personal conversations become, then the more problematic it is to confront realities of exclusion.

5. The local social capital environment is perceived as being a powerful dynamic for change which can successfully challenge institutional and organisational inertia or caution. However, it is recognised that bonding social capital which is internal within organisations, groups and local communities, has the potential to be exclusive, elitist and conservative. Bridging social capital, as a longer term, respect oriented construct, is perceived as a potential modernising influence.

6. There are many organisations with an interest in rural society. These are mindful of the potential which EDI can bring to the achievement of greater social cohesion in Northern Ireland. Much of this work is being done quietly and is targeted at particular groups. Because of its perceived transformational capacity within organisations and across the relationships between organisations and their constituencies, it is regarded as being capable of reaching deeper than the provisions of the statutory equality schemes.

7. EDI is thus perceived as being more than the implementation of the provisions of the equality legislation though it fits well with the 'good relations duty'. It is about trying to encourage people to act in a relational way to others who are different. Organisations are thus challenged to consider not just diversity among rural groupings, but also the particular underlying conditions of inequality which can impact on individuals within these groupings.

8. EDI as a language of diversity and inclusion must intimately connect with local geographies and the many varied interpretations of real and figurative place recognised by rural communities. The deep fractures which exist between the multiple components of rural society are perceived to have spatial consequences in terms of territory and local allegiance.

9. The process is a slow one that needs to be implemented at a pace that the key stakeholders are able to move, especially within their own constituencies; it should not be treated as a mandatory requirement of funding programmes, but as the adoption of core values that are important for their own sake.

10. The ambition of EDI has the potential to disturb or frighten target interests and cause initial rejection of its introduction; this is identified as a key issue when engaging with rural community groups and interest coalitions. EDI discussions must be handled with considerable tact and diplomacy.

11. EDI is not, however, a ready-made product in that there exists one answer and the task is to try to convert others to adopt it. There is the need, rather, to create the space for people to explore their own context, assumptions and relationships and to determine behavioural choices on the basis of that assessment. Different situations, contexts and groups will create different learning experiences. Critical self-assessment, which draws on perception data from organisational constituencies, can enrich this internal analysis.

12. Confidentiality is vital and in this context EDI does not lend itself to early, comfortable and open group discussion. A perceived slight has the grave potential to de-rail a process built around trust, especially in delicately balanced political and ethno-religious contexts. EDI involves taking risks to create confidence about having a dialogue.

13. Champions of the concept should be identified, especially where they share the values of EDI and are suitably connected to, or located within, the relevant organisational hierarchies for initiating, managing and achieving change. Champions also have an important role in marketing EDI, particularly to individuals and groups who share very reasonable dynamics of fear and resistance. In short, the commencement of the process requires committed leadership.

14. The approach must have very senior organisational commitment. It must be enshrined in organisational objectives, the corporate strategy and the review procedures. A written EDI code has both symbolic and practical value.

15. EDI must be able to command political support, especially within local authorities and partnerships. The implications for the way in which political business is conducted is profound and most success can be achieved through intensive working with individual parties and stakeholder groups before reaching across to engage an organisation on a more collective basis.

16. The approach, as demonstrated at District Council level, works best when it is implemented internally before being introduced to external target groups and aimed at external issues. It is worth noting the advances currently being made on this front by the staff of Rural Community Network. It would seem appropriate that at least the bedding down of this initiative should be completed before there is any wider engagement with rural interests.

17. An external input can be vital in securing organisational commitment to a process which is both rigorous and valid. Thus, within Northern Ireland, Counteract and Future Ways, as critical partners in a dialogue process, are perceived to have together developed this necessary practice related expertise to share with participant organisations. But it is also important that some of the language and concepts are translated in a way that can be readily accessible to people with different knowledge levels and experiences.

18. Very focused work on possible outcomes identification, risk assessment, contingency planning and exit strategies needs to be thought through before EDI work is implemented. Nevertheless, precise outcomes are impossible to predict and the process must give space and invitation for any group to move in the direction that is evolving. In other words there must be flexibility within the EDI framework.

19. Constant monitoring, anticipating problems and adjusting programme delivery can ensure that the implementation of EDI is dynamic, responsive and adaptable, especially to unpredictable political events. The evaluation of EDI activities must embrace qualitative measures. However, evaluation has to happen from within the journey of change and not be constrained by artificial indicators established at the start of the process.

20. EDI must be underpinned by specialist training on generic skills for building cross-community respect, including dispute mediation and resolution; this is regarded as being a vital pre-condition before any engagement with rural communities. Sub- regional Rural Support Networks have expressed interest in capacity development and, by reason of their connectedness to localities, can be significant players in this arena of change.

21. And finally, EDI is perceived as being more than a conventional interpretation of community relations which is built around making people and communities apolitical. Thus interviewees frequently expressed their frustration with what are regarded as shallow and 'manipulative' cross-community initiatives. There is a recognised need to commit to long term relational work with single identity groups who may not in the first instance be predisposed to respectful dialogue and behaviour. The enduring challenge, therefore, is to grow people's capacity to deal with complex social, economic, cultural and political tensions regarding how Northern Ireland's rural society organises itself.

PART THREE:
THE WAY FORWARD

Chapter 8

Reconnecting Governance and People Through Authentic Dialogue

Introduction

The focus of this book is on exploring the relationship between public organisations and citizens with a view to better understanding and dealing with the challenges of exclusion, policy responsiveness and societal transformation. The many different realities of rural Northern Ireland provide a useful laboratory to examine difficult questions relating to equity, diversity and interdependence. But that laboratory also throws into sharp relief how it is possible to go beyond minimum standards of practice set down by equality legislation, social inclusion policy and community relations initiatives and move towards a deeper understanding built around mutually respectful inquiry. At a wider level these sentiments have been canvassed by the OECD as constituting the need to harness the capacity of citizens to share with governments a wide sweep of law, policy and institution agenda setting (OECD, 2001). The key driver of that adjustment is authentic dialogue within public organisations, and between public organisations and those in society whom they seek to serve. Our analysis demonstrates that there is no universal road-map and that there are considerable hurdles to overcome. But at the same time we illustrate a combination of belief and possibilities that things can be done differently and better.

Dialogue within the public policy arena has, of course, never been more fashionable. It is not only the quintessential participatory tool of those professionals with an interest in spatial planning, but has an extended reach across all aspects of public sector policy formulation and delivery. The magic word here is 'consultation'. It is appropriate, therefore, that this penultimate chapter should commence by critically reviewing the prominence and promises of a contemporary canon of organisational effort. Out of that assessment emerges the opportunity to reconnect governance and people through a more authentic dialogue that is mutually empowering. The characteristics of and requirements for authentic dialogue are rehearsed and the relationship between authentic dialogue and social capital enrichment within the context of the EDI research reported in the previous section of the book is outlined. The chapter, in short, provides a rationale for systemic adjustment of attitudes, behaviour and perception around the construct of equity, diversity and interdependence.

Consultation and public organisations

The imperative of stakeholder consultation is now at the heart of public policy formation in the United Kingdom. Central government departments, devolved administrations, public agencies and local authorities are all on this treadmill of inviting comment on the content of draft policy papers. Guidelines specify the period of weeks which is desirable for this engagement, outline methodologies for large group interactions, caution against the risks of consultation burn-out, and offer suggestions on how to reach the most excluded or vulnerable people and groups in society. Information technology is advanced as being an appropriate conduit for providing details of existing consultations, notification about forthcoming consultations and the results of completed consultations. A call has even been made for the electronic publication of a comprehensive consultation register to which all UK government departments and agencies would be required to submit details of their consultations in a compatible format (Consultation Institute, 2003). In Northern Ireland, a recently published guide to policy making (Office of the First and Deputy First Minister, 2003) lists ten features of 'good policy making', all of which are denoted by a dependency on consultation (Table 8.1).

Table 8.1 The Ten Features of Good Policy-making

Feature	Consultation dimensions
1. Forward looking	Foresight analysis
2. Outward looking	Comparative analysis of similar issues
3. Innovative, flexible, creative	Process open to suggestions of others
4. Evidence-based	Key stakeholders involved at an early stage; consults with experts
5. Inclusive	Consultation with policy deliverers and recipients
6. Joined-up	Emphasis on negotiating joined-up working arrangements
7. Learns lessons	Emphasis on dissemination
8. Communication	Policy communication with the public
9. Evaluation	Defining success criteria at the outset
10. Review	Service deliverers and recipients provide feedback

Source: adapted from Office of First Minister and Deputy First Minister (2003) A practical guide to policy making in Northern Ireland.

The context presented is one of openness to ideas, a willingness to learn, collective responsibility, and a need for policy messages to be heard. Information gathering, management and dissemination through an array of procedural policy instruments have become crucial elements of managing the 'hollow state' (after Howlett, 2000). This reflects the structural transformation of government behaviour which increasingly obliges it to steer rather than row (Osborne and Gaebler, 1992), to work within a complex framework of policy communities (Rhodes, 1997), to support the management revolution of agencies, charters, market testing and contracting out (Theakston, 1998), and to embrace a multiplicity of multi-organisational partnerships for policy delivery (Lowndes and Skelcher, 1998). In short, consultation is deeply embedded in working through the shift from government to governance whereby the latter can be defined as a process of participation constructed around networks of engagement which attempt to embrace diversity in contemporary society, which promote greater responsiveness to service users, and which seek to reshape accountability relationships (Lovan *et al*, 2004).

Within the sphere of spatial planning there is a long tradition of stakeholder engagement. All too often, however, a professional style which had been fashioned under the hegemony of rational comprehensiveness relegated the role of citizens to a position well below the highly technocratic prescriptions of the planning expert. More than thirty years ago Friedmann (1973) in his transactive model of planning recognised that inputs are required from both technical experts and civil society. The former, he argued, brings valuable theoretical concepts and analysis, processed knowledge, new perspectives, and systematic search procedures for the resolution of problems. Civil society, on the other hand, has a more intimate knowledge of local context and challenges, a better feel for realistic alternatives, and a greater ability to prioritise needs and make feasibility judgements. This participatory approach, premised on interaction, is more demanding than any quick-fix planning solution. But if it leads to a relationship of mutual obligation and reciprocal trust between both parties then a development strategy or plan will be more firmly embedded and stand a healthier chance of successful implementation.

Quality participation is thus contingent on a richness of engagement processes which can transcend consultative tokenism and the ever present risk of cynicism about genuine involvement. This perspective has been examined more recently by Albrechts (2002) who points to the need to innovate through tailor-made approaches that empower participants. He argues that planning cannot be theorised as though its approaches and practices are neutral with respect to class, gender, age, race and ethnicity. This brings attention to the potential contribution of equity, diversity and interdependence for informing both participatory planning theory and practice. As Albrechts states:

> Difference must become a category of analysis within planning theory. This calls for innovations in relations between individuals and between groups in terms of a fair, open and communicative planning process. This involves negotiation and mediation in working through a problem with those directly affected. (Albrechts, 2002, p.332)

In short, there are very striking challenges ahead for the roles and methodologies of consultation within the public policy arena. A provocative analysis by Cornwall (2003) of gender issues within participatory approaches to development provides a very important reminder of how participation and participants are constructed (Table 8.2). The important point here is that there are multiple forms of engagement which vary in the degree to which they genuinely create possibilities for conversation and empowerment. While Arnstein's (1971) ladder of citizen participation made much the same case, the critical contribution of Cornwall is the emphasis given to "combining advocacy to lever open spaces for voice with processes that enable people to recognise and use their agency". It is our contention that these 'spaces for voice' and 'use of agency' can best be located within the authentic dialogue of equity, diversity and interdependence and can shift the practices of consultation from being episodic to continuous, from being linear or sequential to simultaneous, from being formally to informally convened, from its focus on action-oriented decisions to understanding-oriented processes, and from obligations of duty and loyalty to values of trust and reciprocity.

Table 8.2   Modes of Participation

Style of participation	Why invite/involve?	Participants viewed as
Beneficiary	To enlist people in projects or processes, so as to secure compliance, minimise dissent, lend legitimacy	Objects
Community	To make projects or interventions run more efficiently, by enlisting contributions, delegating responsibilities	Instruments
Stakeholder	To get in tune with public views and values, to garner good ideas, to diffuse opposition, to enhance responsiveness	Actors
Citizen	To build political capabilities, critical consciousness and confidence; to enable to demand rights; to enhance accountability	Agents

Source: Cornwall (2003) p.1327

Authentic dialogue

While consultation is commonly regarded, therefore, as the high watermark of dialogue within the public policy arena, the conclusions above suggest that it would seem important to have regard to the style of language being used within any process of exchange. So often the language employed by organisations or those seeking to influence policy agendas is tantamount to rhetoric, with participants adopting a strident adversarialism to represent their views and where the emphasis is on stating a position and defending it. This can be represented, perhaps, as constituting the power of the more attention-grabbing and more emotional argument. Of course where winning is important, rhetoric can be both legitimate and valuable as in many of the social campaign initiatives brought forward into the public arena by civil society. In this sense Aristotle's defence of rhetoric is a valuable reminder of the need for that style of language in advocacy politics:

> even if the true and the just are naturally stronger, avoiding the use of rhetoric, while others do not, weakens your ability to uphold the true and the just. (cited by Remer, 2000, p.90)

The alternative is what Habermas has termed the "force of the better argument". Essentially this requires that participants should present their arguments sincerely, listen to the arguments of others sincerely, and recognise that, even if there is dispute around matters of validity, their dialogue partners are sincere in their desire to reach agreement (cited by Remer, 2000). Participants are thus searching for a common ground through conversational cooperation "which aims at discovering truth, not refuting someone as an opponent" (Remer, 2000, p.74). Such sentiments lie close to the rationale for the empirical research reported in Part Two of this book and throw light on the value of narratives based on personal experience in this exploration of equity, diversity and interdependence. Young (1996) labels this very appropriately as "storytelling" the significance of which strikes a profound resonance with the way that our interviewees gave voice to their very different perspectives. As Young states:

> In situations of conflict that discussion aims to address, groups often begin with misunderstandings or a sense of complete lack of understanding of who their interlocutors are, and a sense that their own needs, desires and motives are not understood. This is especially so where class or culture separates the parties. Doing justice under such circumstances of differences requires recognising the particularity of individuals and groups as much as seeking general interests. Narrative fosters understanding across such difference without making those who are different symmetrical. (p.131)

Young offers three important reasons why narrative matters. First she suggests that narrative "reveals the particular experiences of those in social locations, experiences that cannot be shared by those situated differently but that they must understand in order to do justice to the others" (p131). Accordingly and for example, the qualitative data in Part Two of this book make reference to the plight of Travellers as one of the most socially excluded social groupings in Northern Ireland society. The storytelling from our interviewees identifies problems of access to preventative health care, disinterest by community groups in the cultural heritage of Irish Travellers, the difficulties connected with obtaining planning permission for residential sites, and the need for training within service organisations to transform a perceived institutional insensitivity towards mutual respect. While these experiences cannot be shared they can nevertheless be more fully understood through dialogues of sincerity.

Secondly, Young suggests that our pluralist societies "often face serious divergences in value premises, cultural practices and meanings, and these disparities bring conflict, insensitivity, insult and misunderstanding. Under these circumstances, narrative can serve to explain to outsiders what practices, places or symbols mean to the people who hold them" (p.131). Again our research in relation to the contrasting positions of the Orange Order and the Gaelic Athletic Association fits well with this insight. Each organisation is prominent within rural Northern Ireland and each has its particular practices (parades and games respectively), places (Orange Halls and playing fields respectively), and symbols (the Union Jack flag and the Irish Tricolour flag respectively). Each organisation has a strong cultural affiliation with British identity and Irish identity respectively. The progression of Northern Ireland from a culturally divided society to a more culturally shared society requires that these dialogues of explanation are a first step in building appreciation and respect for positions that are fundamentally disagreed with. At the same time the confidence of participants to share narratives does not require a dilution of the things that they hold dear.

And thirdly, Young suggests that "listeners can learn about how their own position, actions and values appear to others from the stories they tell. Narrative thus exhibits the situated knowledge available of the collective from each perspective, and the combination of narratives from different perspectives produces the collective social wisdom not available from any one position" (p.132). This emphasis on telling and listening is important. In our book we have been careful to describe the institutional or "official discourses" relating to matters of equality, community relations and social need in Northern Ireland. We have also outlined some key characteristics of the rural and the emergence of government responses to conditions of disadvantage and policy neglect. Taken together the narratives which we report from our interviewees constitute a powerful statement of how well the official discourses stack up. It is, perhaps, not surprising that the sincerity of our dialogues allowed for a thus far untold 'collective wisdom' around public policy within the rural. The differential reading of this material by two organisations is interesting. On the one hand, Rural Community Network (the sponsor of the research and whose EDI involvement we detail in Chapter 6) viewed the storytelling as contributing to further political argument around policies and actions. On the other hand, the Department of Agriculture and Rural Development (the lead department for rural affairs in Government) was considerably uncomfortable with the messages being relayed and sought through last minute requests for re-writing and restructuring of the research to both delay and dilute the integrity of the 'situated knowledge available of the collective'. All this is a salutary illustration of the broader proposition that dialogue is never a smooth journey down the path toward agreement or understanding, that the many voices in dialogue are not equal in status, respect or power, and that institutional factors privilege some voices, and silence or discourage others (Burbules, 1993 as cited by Singh, 2001).

In short, authentic dialogue when fully embraced comprises a number of dimensions which can be summarised[1] as follows:

- it is a collaborative process with different interests working together towards a common understanding;
- in dialogue it is necessary to listen to the other sides in order to understand, discover meaning and search for agreement;
- dialogue exposes hidden assumptions and causes reflection on one's own position;
- dialogue has the potential to change a participant's point of view;
- dialogue creates the possibility of reaching better solutions/strategies than any of the original proposals;
- in dialogue participants are asked to temporarily suspend their deepest convictions and to search for strengths in the positions of others;
- dialogue can create a new open-mind set of attitudes: an openness to being wrong and an openness to change;
- dialogue should generate a genuine concern for other persons and thus not seek to offend or alienate;
- dialogue works on the basis that many people have pieces of the desired ways forward and that by talking together they can put them into a workable set of strategies;
- dialogue does not call for conclusions to be reached, but rather is an ongoing process linked with the development of a capacity to think of new strategies and to take sustained actions.

The application of these considerations to equity, diversity and interdependence can ensure that dialogue between service organisations and citizens has the potential to create new contexts, new overarching values and new future directions. Accordingly, authentic dialogue is a voyage of discovery between interlocutors which as Burbules (1993) indicates can give rise to a multiplicity of outcomes:

(1)   agreement and consensus, identifying beliefs or values all parties can agree to;

---

[1] This list is adapted from Berman, S (1996) Paper for Group of the Boston Chapter of Educators for Social Responsibility, cited by Roulier, M (2000) Reconnecting Communities and their schools through authentic dialogue, National Civic Review, Vol 89, No 1, pp.53-65.

(2)   not agreement, but a common understanding in which the parties do not agree, but establish common meanings in which to discuss their differences;

(3)   not a common understanding, but an understanding of differences in which the parties do not entirely bridge these differences, but through analogies of experience or indirect translations can understand at least in part, each others position;

(4)   little understanding, but a respect across differences, in which the parties do not fully understand one another, but by each seeing that the other has a thoughtful, conscientious position, they can come to appreciate and respect even positions they disagree with;

(5)   irreconcilable plurality which is due to participants sharing different cultural values or world-views. (cited by Singh, 2001)

Authentic dialogue in the context of EDI is not, therefore, a one shot venture. Nor should authentic dialogue within this sphere be perceived as a hierarchical construct where the best outcome is agreement and consensus and the worst is irreconcilable plurality. The listing advanced above by Burbules does indicate a capacity for differential outcomes. However, within our analysis of EDI what is important is that conversations not only take place but that conversations continue to take place. As we indicate in Chapter 7, the EDI process is a slow one that needs to be implemented at a pace that the key stakeholders are able to move, especially within their own constituencies, that the ambition of EDI has the potential to frighten target interests and cause initial rejection, and that EDI is not a ready-made product with one answer and where the challenge is to try to convert others to adopt it. There is a need to create the space for people to explore their own context, assumptions and relationships and to determine behavioural choices on the basis of that assessment. Different situations, contexts and groups will generate different learning experiences.

The research reported in Part Two of this book demonstrates time and time again the need for and value of a concerted approach to authentic dialogue in order to facilitate systemic relational change within organisations as well as between organisations and their service constituencies. Dialogue allows citizens holding diverse perspectives to engage with each other in inquiry, sharing and learning. It allows service organisations and those people with whom they interact to begin to discover principles for a more productive engagement out of which can flow new practices of understanding and respect. When embedded in behaviour as everyday efforts the consequences for social capital enrichment can indeed be extremely powerful.

Authentic dialogue and social capital

Social capital describes the resources which are stored in personal relationships, whether casual or close (de Souza Briggs, 1997: 112). It does not equate simply with civic engagement, though participation in community life by, for example, committing time to the work of a local development association, can help to generate social capital by bringing people into contact with each other. Social capital facilitates collective action based on trust and thus nurtures an ability to collaborate for shared interests. Robert Putnam (1993) in his seminal book *Making Democracy Work* expresses this in a more prosaic form: 'I'll do this for you now knowing that somewhere down the road you'll do something for me' (p.182). He argues that without these norms of local reciprocity and networks of engagement the outlook in any society is bleak. Outcomes include clientelism, lawlessness, economic stagnation, and ineffective government. Thus strong participatory citizenship whereby people are involved in planning and in implementation, in facilitative leadership roles and in creating better futures for their own organisations or communities is inextricably linked to the existence of social capital networks of engagement.

Social capital is an important relational asset and as noted by Flora (1997) the networks creating social capital are multi-dimensional. The configuration encompasses lateral learning both between and within organisations and communities, linking the local more closely to regional and national resources and organisations, allowing people to engage and disengage over time, and accepting that community and organisational collaborations and interdependencies can expand and narrow depending on the issues being addressed.

Foley and Edwards (1999) express this as the need to distinguish between the possession of social capital by individuals and collective actors and its varying use in a manner which takes account of context dependency. Thus, social capital does not equate simply with informal goodwill. Lappe and Du Bois (1997) caution that even though people can find that level of engagement enriching, these networks of association can still co-exist with a generalised sense of powerlessness. In this context they advocate that building social capital rests on a foundation of three requirements: (1) a sense of hope by people that solutions are possible; (2) sufficient opportunities for engagement by those with the necessary motivation and skills; and (3) opportunities to nurture service to society life-skills. Ultimately their presence is conditional upon greater trust and less regulation by public officials and senior managers and their willingness to support a 'let the flowers bloom' strategy of relationship building. It is a truism that the more social capital is invested in, the greater is the capacity of an organisation to achieve credible product outcomes. At a time when public policy is universally obsessed with performance and results, investing in social capital processes can easily be tarnished by the superficiality of a solely value-for-money judgement. But, it is people, not organisations per se, with their beliefs, attitudes and actions that can make a difference and thus the significance of social capital formation deserves a better fate.

The literature highlights that social capital has three components (Morrissey *et al*, 2002):

- bonding social capital which relates to internal cohesion or connectedness within a group;
- bridging social capital which relates to the levels and nature of engagement with other groups;
- linking social capital which comprises the relationships, for example, between civil society and service organisations and policy makers.

Each component is moderated by interactions with the other two and thus all three require simultaneous nurturing. This point can be illustrated by the reality of paramilitarism as a malign form of bonding social capital and where organisational capacity, leadership and community connectedness are directed toward anti-social purposes. As noted by Putnam (2000) "networks and the associated norms of reciprocity are generally good for those inside the network, but the external effects of social capital are by no means always positive" (p.21). It is our contention that authentic dialogue around matters of equity, diversity and interdependence can provide the necessary glue to bind together these three dimensions of social capital thus maximising its positive consequences. Our empirical research in Part Two of this book provides useful insight around this observation.

In Chapter 6 we detail the efforts being made by the membership organisation, Rural Community Network, to more fully understand how it is perceived externally. The staff and management of this organisation have sought to transform the established internal environment into a new social context of shared learning based on dialogue. There is enthusiasm and commitment that this should continue. The bonding social capital outcomes are evidence, therefore, of mutually beneficial rewards for the sincerity of that dialogue. But at the same time, Rural Community Network is appreciative of its responsibility to transmit knowledge, skills and values around equity, diversity and interdependence to a large number of community groups and their volunteers. The publication of a workbook and training seminars are viewed by Rural Community Network as an important way to confront rural communities with the need to accept and manage change. The crucial linkage here is not that between Rural Community Network and these groups, but rather between the many community groups with their often contrasting cultural identities operating in often contested local territories. Bridging social capital when fostered through authentic dialogue widens awareness of the many ways in which local situations are linked to common causes and allows citizens to resolve collective problems more easily through collaborative initiative (after Putnam, 2000, p.288).

Finally, it is important that mention is made of the linking social capital dimension which defines the relationship between the many region-wide and area-based service organisations we interviewed and rural people. It is a relationship built around resources (funding, service, representation) and involves disbursement of resources by the former and access to resources by the latter. Our dialogues capture serious misperceptions, the dilemmas of difficult issues to talk about, a combination of organisational neglect, minimalism and blindness, as well as powerful illustrations of exploration and learning. As lived experiences, our interviewees have expressed their frustrations, hurts, and hopes with conviction and honesty. Authentic dialogue constructed around equity, diversity and interdependence has thus provided a necessary space for critical reflection on matters of policy content, policy delivery and policy beneficiaries. Our research provides evidence of the transformative capacity of EDI and also signposts how best the approach should be initiated within any organisational setting. The important point here is that this investment into linking social capital through authentic dialogue can have significant consequences for organisational effectiveness.

Conclusion

The author Anna Funder in her much acclaimed 2003 book *Stasiland* presents a suite of stories around how the well perfected surveillance state of the former East Germany impacted on the lives of ordinary people. The narratives draw on the experiences of former Stasi officials as well as those whom the state persecuted. The system depended on dialogue:

> It was a bureaucracy metastasised through East German society: overt or covert, there was someone reporting to the Stasi on their fellows and friends in every school, every factory, every apartment block, every pub. (p.5)

Such darkness throws our representation of authentic dialogue into sharp relief. It is a dialogue which, in contrast, seeks to rekindle relationships of trust and respect between governance and people at a time when there is considerable public cynicism for those in charge and yet when those in charge place great weight on public consultation. Our argument is that sincere conversations constructed around equity, diversity and interdependence can provide an important antidote to the distrust and indifference of citizens, while also invigorating the processes of public policy formulation and delivery. The laboratory of rural Northern Ireland demonstrates scope for the wider adoption of deeper interaction between public organisations and those whom they seek to serve thus bridging the often yawning gap between official policy discourses and what things are really like 'on the ground'. In Chapter 1 we signalled the relevance of this approach for re-working ideas around collaborative planning. It is appropriate that our final chapter should more fully respond to that challenge.

Chapter 9

Collaborative Planning in Action

Introduction

The previous chapter demonstrated the prescriptive value of dialogue for community development and practice and in this chapter we reflect on the theoretical implications of the research. In the opening chapter of the book we described the origins of collaborative planning, its application to highly ethnicised places and the way in which equity, diversity and interdependence are recurring themes in a number of societies emerging from conflict. However, where ethnic divisions are overlain by a range of exclusions linked to gender, disability, age, race and sexual orientation, the challenge to planning and development is intensified. Moreover, where these divisions are increasingly spatially correlated, such as in the rural areas of Northern Ireland, 'wicked problems' represent formidable obstacles to local change and growth. The theoretical questions posed at the start of this book centred on the analytical and prescriptive capacity of collaborative planning to help direct rural development in these places and problem contexts. The collaborative planning model was developed, and largely applied in the Western English-speaking world but it assumes much about political cultures, institutional capacities and policy styles to operate via a vocabulary in which the 'power of the better argument' would produce consensual planning outcomes. We have seen that, in the polarised and highly segmented world of rural Northern Ireland, the notion of consensus, especially over territory, is difficult to identify let alone achieve.

Agonism

Hillier (2003) has offered an alternative agenda for planning practice based on the notion of *agonism* and the healthy political effects of discord in communities attempting to resist the consequences of global and industrial change. We saw earlier that Parker (2001) identified 'sites of resistance' where racially and socially excluded communities occupy the most marginal places in the post-modern city. There is potential in these places because this concentrated discontent can be manipulated to achieve political change and better conditions for local people. Radical geographers such as Soja, Healey and Castelles view this as redefining spatial studies, the policy agenda and planning specifically (see Soja, 2000). However, these accounts often fail to define, with any clarity, how that agenda might be taken forward in specific situations or how planners or rural activists could respond in their own practice areas. This is precisely where Equity Diversity and Interdependence (EDI) has something to offer as it can also help to animate collaborative planning in places divided by poverty, social alienation and religion.

Here, it can help to address some of the substantive limitations of collaboration by giving purpose to discourse and making it relevant to specific outcomes of dialogic processes. Collaboration works to achieve what Healey (2003) terms 'place quality' but EDI assumes that places cannot be objectivised in this way – they are not the outcome of a technical rational system – but instead shape the conversations, what is possible and what is not and suggest the type of social spaces the planning process might work towards. In this case, EDI helps us to think about communities that are less socially divisive, where identity is respected but where interdependent alliances between marginal interests can create new possibilities for action. In short, EDI can give substantive direction to the discourses essential in creating more equal, diverse and interdependent places. Discourse cannot be treated in a detached rational way but is framed around specific planning outcomes. Space cannot be objectivised but is imbued with social and political significance, which itself, influences the range and depth of the discourse and what it is likely to achieve in reality.

The research highlighted the 'dark side of difference', the prevalence of agony and the limits on discourse in the particular cultural and political context of rural society in Northern Ireland. Agonism is not new and its destructive qualities have been identified in the empirical research. We saw in the last chapter that social capital is a valuable asset in helping rural communities to develop indigenous capacities, economic development projects and even local services. Yet, the agonism between cultural institutions such as the Orange Order and the GAA, highlight the possibility for conflict between different identities. Equally, poor people can activate around mutually exclusive and antipathetic networks, structures and activities, which reduces the prospect for a meaningful discourse about poverty, the effects of agricultural restructuring and the possibilities for locally based, community-led change. Spatial and cultural segregation has long been a logical response to conflict, fear and the need for safety (Murtagh, 2002). However, whose interests are now best served by continuing socio-spatial segregation and who loses from it are interesting questions addressed in the research here. In particular, EDI focuses attention on the cultural institutions that play a role in the reproduction of exclusive identities and, at the same time, build capital around which communities can organise to achieve change.

However, by widening the identification of stakeholders to embrace a richer set of connected interests, a broader picture has emerged of rural life and development issues, such as, for example, Protestant alienation and decline. In some places, the Protestant population has fallen and its institutional capacity (including schools, churches and community halls) has been eroded. Young people and families move out and older people lose valuable support networks, accelerating the residualisation of the local population. It is the younger, employed and employable who move, leaving behind an older, less mobile and benefit-dependent population. The interview data from the Orange Order demonstrates the capacity of that body to organise around a more positive set of local planning objectives. The weaker community capacity of the Protestant community is recognised and the need to assemble and broaden the infrastructure that could help to develop projects, draw down grants and target vulnerable communities are now reflected in a new concern for community development. The organisation's human and physical asset base is being deployed in a positive expression of local development and dialogue. Moreover, the organisation's traditional identity as a negative and exclusive entity has been challenged by its capacity to entertain a new set of relationships in rural society. The Orange Order has not embraced a cross-community curriculum and is still nervous of the wider reconciliation agenda, but in the context of Northern Ireland, the shift from a purely cultural to a more developmental agenda is significant and the capacity to further extend its reach is one of the key challenges that discursive-led EDI framework could help to meet.

Connecting agendas

While it is important not to overstate the experiences of the Orange Order, the case study and the wider empirical research highlight the value in connecting identity and social need issues. In Northern Ireland there is a strong statutory and policy framework upon which equality and Targeting Social Need (TSN) can reinforce each other on a spatial basis. Yet, we have seen from the empirical research that TSN has been interpreted in a formalistic, top-down way. Compliance with the legal minimum rather than stretching the potential of each agenda and capturing the synergies between them seem to have characterised Government approaches. Technocratic policy styles have helped to insulate the local state from accusations of discrimination and bias, especially after the imposition of Direct Rule in 1972. In the pre-1972 administration, major allocation decisions in strategic planning, housing and the treatment of rural affairs highlighted the sectarian nature of the local state and the manipulation of the land use planning system in particular. In order to 'steer' the state through this crisis, new centralised bureaucracies were established, policy-making systems were made highly regimented and professional values were prioritised in key policy areas including planning, housing and rural development. Subsequently, the Good Friday Agreement and key legislative developments in human rights, equality and social justice faced considerable institutional obstacles when it came to policy development and implementation. A collaborative interpretation of these changes helps to pinpoint variations in the policy styles adopted at different times by different organisations. EDI emphasises a more interpretative style of analysis and decision-making where there is scope for diagnosis of different positions, uncertainty and flexibility. Our research shows that many respondents feel that the real value of EDI is to help shift policy makers to a truer engagement of the meaning of equality and Targeting Social Need. In particular, it carries this through by identifying the key interests in local development, building a more detailed understanding of their problems and priorities, exploring overlaps and connections between interests and seeing where there are issues that cannot be connected, at least in the short term.

But, collaborative planning analysis also identifies resistance to issues around religion and segregation in the community and voluntary sector. It shows for instance, that some community groups have found ways of communicating across ethnic-religious and, indeed, other social divides by 'parking' sensitive issues likely to create discord and division. In one sense, groups have become adept at carefully negotiating a form of discourse that can agree agendas where progress can be made and to intuitively define areas where it cannot. This means, however, that many issues crucial to long-term sustainable development have been hidden, denied or routinised. There is a measure of caution and fear that is mixed with negative experiences where explicit discourses around cross-community identity have damaged contact and progress on development issues. We saw, for example, from one case, how a group believed that many problems around the conflict had been raised, but that it had no real mechanism for resolving or developing an analysis of the difficult issues that were created by this more open discourse. All of this highlights the problematic nature of discourse itself and some of the simplistic notions that collaborative planning makes about the type and quality of the conversations that are possible in highly charged and politicised contexts. Moreover, it demonstrates the need for skills in constructing and managing discourse in those situations where the risk of failure or the creation of unintended side effects are a very real prospect.

It is our view that essential competencies for the planning profession, local community groups and development interests comprise mediating disputes, identifying legitimate interests and their concerns, brokering agreements and describing contingencies when agreement cannot be reached. Professional institutes have identified negotiating skills as a key component of practice, yet they say little about how this element might underpin collaborative strategies, especially in divided places where discursive traditions are not well developed. It is crucial that these 'ways of knowing' and the ethnographic tradition are given space in the curriculum of planning schools, particularly since positivist values are by tradition prioritised in professional training and practice.

In short, collaborative planning *per se* has limitations, especially for the way in which it objectivises space and assumes that 'place qualities' can be achieved via dialogue between consensual parties. The reality is that there are cultural and institutional obstacles to contact. Developing trust, working intensively on a pre-dialogue phase to allow later conversations to happen and to supporting people with the necessary skills to sustain tenuous contact are valuable in securing longer-term societal transitions. This is especially the case in the context of 'interdependence' around issues of common concern including poverty, isolation and alienation. Community transport, for instance, meets the needs of disabled people, older people, young people and car-less rural dwellers. Seeing the relevance of their connectedness in particular places and animating it via discursive practice holds important potential for the wider regeneration agenda. In the same vein, links across ethno-religious divides are fragile and are easily checked by the micro-politics of sectarianism and paramilitarism. But, "there is no formula here other than the engineering of endless tasks of interaction between adversaries or provision for individuals to broaden their horizons, because any intervention needs to work through, and is only meaningful in a situated dynamic" (Amin, 2002, p.969). That planning practice might provide that dynamic represents the potential of the EDI agenda, particularly where it can be grounded in the lived experiences of marginal people in divided places.

Conclusion

Equity, diversity and interdependence (EDI) are terms, concepts and aspirations. They are lenses to help interpret society, both rural and urban. They provide the potential for discourse to help achieve sustainable communities. Territorial conflicts bring a different meaning to 'place' and they cannot be treated improblematically in discussion about land use change. It is not surprising, therefore, that ownership, control, retrenchment, loss and appropriation are all part of the vocabulary of rural development and planning in Northern Ireland. The value of collaborative planning, when linked with an appreciation for EDI, is that it can help to expose the nature of these contexts, identify where alliances are possible and envision how fragile connections might be supported and ultimately strengthened through dialogue and trust. In particular, it can help to identify the way in which marginalized citizens, older people, disabled people, ethnic minorities or younger people can form links, which cut across atavistic allegiances in particular places. This is a prescriptive agenda that has implications well beyond rural Northern Ireland, not least in other societies struggling to come out of conflict or to come to terms with racial and social separation.

Glossary

Best Value

Best Value is the way UK local government measures, manages and improves its performance. Best Value can apply to a service, a department or a complete council where the overall objective is to deliver better and more responsive public services. This is achieved by securing a balance between quality and cost. It requires an effective dialogue with local communities to ensure accountability.

Central Community Relations Unit (CCRU)

The Central Community Relations Unit was established in 1997 to advise the Secretary of State for Northern Ireland on all aspects of the relationship between the different components of the local community. The Unit was charged with policy formulation and review within the areas of equality and community relations. In 2000 CCRU was renamed the Community Relations Unit and became part of the Equality Unit of the Office of the First Minister and the Deputy First Minister.

Commission for Racial Equality (CRE)

The Commission for Racial Equality is a publicly funded, non-governmental body set up under the Race Relations Act 1976, to tackle racial discrimination and to promote racial equality. It provides information and advice to people who think they have suffered racial discrimination or harassment, and works with organisations across all sectors to promote policies and practices that will help to ensure equal treatment for all. This legislation applies only to England, Scotland and Wales.

Community Relations Council (CRC)

The CRC was formed in 1990 as an independent company and registered charity, having originated in 1986 as a proposal of a research report commissioned by the Northern Ireland Standing Advisory Committee on Human Rights. Its brief is to promote better community relations between Protestants and Catholics in Northern Ireland, and to promote cultural diversity.

Community Relations Training and Learning Consortium (CRTLC)

CRTLC is a membership based organisation committed to promoting quality community relations learning experiences on the challenges of building an inclusive and plural society in Northern Ireland.

Counteract

Counteract is an anti-intimidation unit in Northern Ireland which was formed in 1991 with the support of the Irish Congress of Trade Unions. Its purpose is to develop actions, policies and strategies that alleviate the incidence of sectarianism and intimidation in the workplace and the wider community. Counteract has been working to facilitate the embedding of internal capacity within organisations regarding the promotion of equity and the acceptance of diversity.

Department of Agriculture and Rural Development (DARD)

The responsibilities of DARD include the development of the agricultural, forestry and fishing industries of Northern Ireland, rural development in Northern Ireland, providing an advisory service to farmers, agricultural research and education, animal health and welfare, and the application of UK and EU schemes and policy within Northern Ireland.

Direct Rule

The imposition of Direct Rule is very much a product of the political turmoil in Northern Ireland from the late 1960s. It was imposed under the Northern Ireland (Temporary Provisions) Act 1972 following the resignation of the Northern Ireland Government and led the United Kingdom Government to assume full and direct responsibility for the administration of Northern Ireland under the then newly created post of Secretary of State for Northern Ireland. Direct Rule has continued during those periods when the Northern Ireland Assembly (established under the Good Friday Agreement) has been in suspension.

District Council

Local government in Northern Ireland comprises 26 District Councils. Compared with local authorities in the remainder of the UK they have limited functions. These comprise direct service provision (refuse collection and disposal, leisure and community services, street cleaning) and regulatory oversight (building control and environmental health). They also have a limited role in economic development and community relations services. Their representative role extends to membership by nominated councillors of a range of statutory bodies. Finally, they have a consultative role, notably in relation to development control and development plan preparation activities. Responsibility for the latter rests with the Planning Service of the Department of the Environment for Northern Ireland.

District Partnership

The initial implementation of the EUSSPPR, following its approval in 1995 by the European Commission, was partly through the creation of 26 District Partnerships (one in each of the 26 District Council areas). They comprised representatives drawn from local government, the voluntary and community sectors, trade unions, business interests and public bodies. A prominent element of their expenditure comprised tackling social exclusion. District Partnerships were replaced by Local Strategy Partnerships in 2002.

Equality Commission for Northern Ireland

The Race Relations Act (Northern Ireland) 1997, based on comparable 1976 legislation in Great Britain, resulted in the establishment of a separate Commission for Racial Equality for Northern Ireland. Under the Northern Ireland Act 1998 the Commission for Racial Equality for Northern Ireland was merged with the Fair Employment Commission, the Equal Opportunities Commission for Northern Ireland and the Disability Council to form a new Equality Commission for Northern Ireland. The Equality Commission is charged with working towards the elimination of discrimination, promoting equality of opportunity and good practice, promoting good relations between people of different racial groups, and overseeing the application of the statutory duty on public bodies.

Equality Impact Assessment (EQIA)

Northern Ireland is unique in pioneering Equality Impact Assessments across a much broader and more inclusive range of categories than any other jurisdiction. An EQIA comprises a systemic analysis of a policy with reference to the nine equality categories as defined by Section 75 of the Northern Ireland Act 1998. The primary function of the EQIA approach is to determine the extent of differential impact of a policy upon the groups and subsequently whether that impact is adverse by having a negative impact on groups in relation to one or more of the nine equality categories. If it is decided that the policy impact is negative, the public authority must consider measures which might mitigate the adverse impact and alternative policies which might better achieve the promotion of equality of opportunity. EQIAs require the analysis of quantitative and qualitative data.

Equality Schemes

Equality Schemes are intended to change the culture of public decision making by placing a more proactive approach to the promotion of equality at the heart of public policy. More particularly, Equality Schemes show how a public body proposes to fulfil the duties imposed by Section 75 of the Northern Ireland Act 1998 which obliges public authorities to have due regard to the need to promote equality of opportunity between nine different categories of people defined by religious belief, political opinion, racial group, age, sexual orientation, marital status, gender, disability and dependency. An Equality Scheme will set out the public authorities arrangements for carrying out consultations, policy review and selection for full equality impact assessments, monitoring, training and access to information.

European Union Special Support Programmed for Peace and Reconciliation in Northern Ireland and the Border Counties of Ireland (EUSSPPR)

The EUSSPPR was approved by the European Commission in July 1995 and was designed initially to reinforce progress towards a peaceful and stable society and to promote reconciliation by increasing economic development and employment, urban and rural regeneration, social inclusion and cross-border cooperation.

Future Ways

The Future Ways Programme is an initiative based in the University of Ulster, Northern Ireland, which seeks to bridge the gap between the history of conflict handling work in the community and voluntary sectors and the need to support practical developments within a large number of institutions and organisations within the region. A key part its work is developing and delivering training on issues of diversity, trust building and equity in different learning contexts.

Gaelic Athletic Association (GAA)

The GAA is an amateur sporting organisation founded in 1884 to preserve and cultivate the national games of Ireland. It is the largest sporting organisation in Ireland with 2,800 clubs and some 280,000 players. The Association is nationalist in outlook.

Good Friday Agreement

The Good Friday Agreement (also called the Belfast Agreement) was reached following intensive multi-party talks on 10th April 1998 (Good Friday). It was subsequently submitted to a referendum in both parts of Ireland on 22nd May 1998. Its content includes matters relating to the position of Northern Ireland within the United Kingdom, devolution of a wide range of executive and legislative powers to a Northern Ireland Assembly, a North/South Ministerial Council, a British – Irish Council, and a new British – Irish Intergovernmental Conference.

International Fund for Ireland (IFI)

The IFI was established as an independent international organisation by the British and Irish Governments in 1986. With contributions from the United States, the European Union, Canada, Australia and New Zealand, the objectives of the Fund are to promote economic and social advance; and to encourage contact, dialogue and reconciliation between nationalists and unionists throughout Ireland.

LEADER

LEADER was introduced as a Community Initiative of the European Union in 1991 and has over the intervening period been concerned with innovation through the development of local endogenous potential; it is area based, bottom-up, partnership based, multi-sectoral, decentralised and supportive of networking. LEADER 2 was designed as continuation of LEADER 1 and the current programme known as LEADER+ runs through to 2006.

LEADER 2 Local Action Groups (LAG)

LAG are a partnership based mechanism for the delivery of LEADER at the local level.

Local Strategy Partnerships (LSP)

LSP are located within each of the 26 District Council areas of Northern Ireland. They are responsible for implementing two Measures under Priority 3 of the EU Programme for Peace and Reconciliation 2: Measure 1 – local economic initiatives for developing the social economy, and Measure 2 – locally based human resource, training and development strategies. Their memberships draw from a combination District Council, voluntary, community, trade union, business and statutory agency sectors.

Mediation Network

The Mediation Network promotes the use of Third Party intervention in disputes and supports creative responses to conflict in Northern Ireland. It is staffed by a cadre of mediation practitioners, provides training in community relations, conflict intervention and mediation, and provides mediative assistance in disputes.

New TSN

This aims to tackle social need and social exclusion by targeting efforts and resources towards people, groups and areas in greatest social need. It is a principle which runs through the spending programmes of Northern Ireland Government departments. It is aimed at tackling unemployment and enhancing employability, tackling inequality in other policy areas (e.g. health, housing, education), and promoting social inclusion through collaborative working with others inside and outside Government.

Northern Ireland Agricultural Producers Association (NIAPA)

NIAPA is a representative organisation of hill farming interests in the Less Favoured Areas of Northern Ireland. At any point in time it has between 4000 and 5000 members engaged primarily in beef and sheep production.

Northern Ireland Housing Executive (NIHE)

The NIHE is the regional housing authority for Northern Ireland and has responsibility for examining housing conditions and housing requirements, dealing with unfit houses, encouraging the provision of new houses, and managing its own housing stock.

Northern Ireland Rural Development Programme (RDP)

The Northern Ireland Rural Development Programme seeks to bring help and practical support to people living in Northern Ireland's rural communities. The current phase runs from 2001-2006 and is concerned with building the capacity of rural communities, local regeneration projects and programmes, sectoral and area regeneration projects and programmes, natural resource rural tourism and micro business development. The Department of Agriculture and Rural Development (DARD) manages the programme in partnership with Rural Community Network (RCN) and the Rural Development Council (RDC).

Northern Ireland Voluntary Trust (NIVT)

NIVT is an independent charitable grant-making organisation whose aim is to create a more just and caring society. It funds community development and assists groups to tackle the causes and effects of inequality, poverty and disadvantage at local and regional levels. In 2002 NIVT changed its name to The Community Foundation for Northern Ireland. It works across all the divisions within society in Northern Ireland.

Open College Network

The Northern Ireland Open College Network offers support, training and consultation services to providers of education and training and seeks to promote wider access to learning. It is a federation of providers spanning the statutory, community, voluntary, trade union and private sectors. The Open College Network oversees an accreditation process of learning programmes whose standards are consistent across the UK.

Orange Order

The Loyal Orange Institution (more commonly referred to as the Orange Order) is a Protestant fraternity with members throughout the world. Its name honours William 111, Prince of Orange. It is primarily a religious organisation but also accepts its political responsibilities. The Grand Orange Lodge of Ireland is based in Belfast.

Playboard

Playboard is the lead agency for children's play in Northern Ireland, working to improve the quality of children's lives by increasing their opportunity to play. Its activities include giving support to local communities in developing quality play provision.

Policy Appraisal and Fair Treatment (PAFT)

The PAFT Guidelines were the first non-statutory attempt at mainstreaming equality in Northern Ireland. Their aim was to ensure that issues of equality and equity informed public policy making in all spheres and at all levels of government. PAFT is superseded by Section 75 of the Northern Ireland Act 1998.

Rural Community Network (RCN)

RCN is a voluntary organisation established in 1991 by local community organisations in Northern Ireland to articulate the voice of rural communities on issues relating to poverty, disadvantage and equality. It is a membership organisation with over 500 members.

Rural Development Council (RDC)

RDC was established in 1991 under the Department of Agriculture's Rural Development Programme. It exists to address the needs of deprived rural areas and works in partnership with a wide spectrum of interest groups.

Rural proofing

Rural proofing is a commitment by the UK Government to ensure that all its domestic policies take account of rural circumstances and needs. It is a mandatory part of policy design and delivery processes.

Standing Advisory Commission on Human Rights (SACHR)

SACHR has now been replaced by the Northern Ireland Human Rights Commission. As part of its review of mechanisms in place to promote employment equality and reduce the employment differential SACHR recommended that the PAFT guidelines should be made a statutory requirement.

Sub-Regional Rural Support Networks (RSN)

There are 12 Sub-Regional Rural Support Networks across Northern Ireland. Each is a fully constituted non-profit taking organisation with charitable aims and is managed by a committee drawn from local rural communities. Each provides development, technical and support services and all are seen as essential to increasing the diversity of existing groups in rural communities and to enabling marginalized rural dwellers take a more active role in changing their own locality. Support to the RSN is provided by Rural Community Network.

The Special European Programmes Body

This is one of six cross-border Bodies established under the international agreement between the Governments of the UK and Ireland signed on Good Friday 1998. Its brief includes acting as the managing authority for PEACE 2 and INTERREG 3A, along with oversight of the cross-border elements of other Community Initiatives (LEADER+, URBAN 2, and EQUAL. It reports to the North/South Ministerial Council and is accountable to the Northern Ireland Assembly and the Oireachtas.

The Women's Institute

This is an organisation for women across the United Kingdom that exists to educate women to enable them to provide an effective role in the community, to expand their horizons, and to develop and pass on important skills. It was formed in 1915 and works to ideals of truth, justice, tolerance and fellowship.

Traveller Movement (Northern Ireland)

Traveller Movement (Northern Ireland) was founded in 1981 as the Northern Ireland Council for Travelling People. Today it is an umbrella group comprising Travellers, Traveller Support Groups and statutory and voluntary groups. Its aim is to work for justice, equality and equity for Travellers in partnership with others in the fields of education, accommodation, health, economic regeneration and training.

Ulster Farmers Union (UFU)

The UFU is a democratic voluntary organisation representing farmers and growers in Northern Ireland. It takes a close interest in rural affairs and works with politicians within the UK and internationally, together with other organisations to advance rural interests.

Youth Action Northern Ireland (YANI)

YANI is a voluntary youth organisation, which since its inception in the 1940s has sought to make a contribution to the lives of young people by stressing the importance of personal, creative, social and political development. Its work spans community relations, gender equality, working with young men, and rural development.

References

Ahmed, Y. and Booth, C. (1994) Race and planning in Sheffield, in H. Thomas and
V. Krisharayan (eds.), *Race and planning: policies and procedures*, Aldershot:
Avebury.

Albrechts, L. (2002) The planning community reflects on enhancing public
involvement: views from academics and reflective practitioners, *Planning
Theory and Practice*, Vol. 3, No. 3, pp. 331-347.

Amin, A. (2002) Ethnicity and the multicultural city: living with diversity,
Environment and Planning A, Vol. 34, No. 1, pp. 959-980.

Armstrong, J., Mc Clelland, D. and O'Brien, T. (1980) *A policy for rural problem
areas in Northern Ireland: a discussion document*, Belfast: Ulster Polytechnic.

Arnstein, S (1971) A ladder of citizen's participation, *Journal of the American
Institute of Planners*, Vol. 35, No. 7, pp. 216-224.

Atkinson, R. (1999) Countering urban social exclusion: the role of community
participation in urban regeneration, in Haughton, G. (ed.) *Community economic
development*, pp. 65-78, London: The Stationery Office.

Berman, S. (1996) Paper for Group of the Boston Chapter of Educators for Social
Responsibility, cited in Roulier, M. (2000) Reconnecting communities and their
schools through authentic dialogue, *National Civic Review*, Vol. 89, No. 1, pp.
53-65.

Birrell, D. and Murie, A. (1980) *Policy and government in Northern Ireland:
lessons of devolution*, Dublin: Gill and Macmillan.

Bollens, S. (1999) *Urban peace-building in divided societies: Belfast and
Johannesburg*, Boulder, CO: Westview Press.

Bollens, S. (2002) Urban planning and intergroup conflict: confronting a fractured
public interest, *Journal of the American Planning Association*, Vol. 68, No. 1,
pp. 22-42.

Booher, D. and Innes, J. (2002) Network power in collaborative planning, *Journal
of Planning Education and Research*, Vol. 21, pp. 221-236.

Brownhill, S., Razzaque, K., Stirling, T. and Thomas, T. (1997) Local governance
and the racialisation of urban policy in the UK: the case of Urban Development
Corporations, *Urban Studies*, Vol. 33, No. 8, pp. 1337-1355.

Bryan, D. (2000) *Orange parades: the politics of ritual, tradition and control*,
London: Pluto Press.

Burbules, N.C. (1993) *Dialogue in teaching: theory and practice*, New York:
Columbia University.

Caldwell, J. and Greer, J. (1983) *Planning for peripheral rural areas: the effects of
policy change in Northern Ireland*, Social Science Research Council Final
Report, Belfast, The Queen's University of Belfast.

Christopher, A. J. (1998) (De)segregation and (dis)integration in South African metropolises, in Musterd, S. and Ostendorf, W. (eds.) *Urban segregation and the welfare state,* pp. 227-241, London: Routledge.

Cloke, P. (1978) Changing patterns of urbanisation in the rural areas of England and Wales 1961-1971, *Regional Studies,* Vol. 12, pp. 603-617.

Community Relations Council (1998) *Into the mainstream: Strategic Plan 1998-2001,* Belfast: Community Relations Council.

Connolly, P. (1998) *Early years anti-sectarian television,* Belfast: Community Relations Council.

Consultation Institute (2003) *The Consultation Institute's preliminary response to the Draft Code of Practice on Consultation,* Orpington: The Consultation Institute.

Cornwall, A. (2003) Whose voices? Whose choices? Reflections on gender and participatory development, *World Development,* Vol. 31, No. 8, pp. 1325-1342.

Counteract and Future Ways Programme (2001) *Equity, Diversity, Interdependence - Gaining from difference: a framework for change,* Coleraine: University of Ulster.

Cousins, C. (1998) Social exclusion in Europe: paradigms of social disadvantage in Germany, Spain, Sweden and the United Kingdom, *Policy and Politics,* Vol. 26, No. 2, pp. 127-146.

Department for Regional Development (2001) *Shaping our future: the Northern Ireland Regional Development Strategy,* Belfast: The Stationery Office.

Department of Agriculture and Rural Development (2001) *The Rural Development Programme 2001-2006,* Belfast: Department of Agriculture for Northern Ireland.

Department of Agriculture for Northern Ireland (1999) *The Rural Development Programme in Northern Ireland 1994-1999 - progress review,* Belfast: Department of Agriculture for Northern Ireland.

Department of the Environment for Northern Ireland (1993) *A Planning Strategy for Rural Northern Ireland,* Belfast: The Stationery Office.

Department of the Environment for Northern Ireland (1998) *Draft regional strategic framework for Northern Ireland,* Belfast: The Stationery Office.

Department of the Environment for Northern Ireland (1998) *Shaping our future: the family of settlements report,* Belfast: The Stationery Office.

Department of the Environment, Transport and the Regions (DETR) (1997) *Single Regeneration Budget Challenge Fund Round 4: Supplementary Guidance,* London: DETR.

De Souza Briggs, X. (1997) Social capital and the cities: advice to change agents, *National Civic Review,* Vol. 86, No. 2, pp. 111-117.

Dorsett, R. (1998) *Ethnic minorities in the inner-city,* Bristol: The Policy Press.

Duffy, P. (1997) Writing Ireland: literature and art in the representation of Irish place, in Graham, B. (ed.) *In search of Ireland,* pp. 64-83, London: Routledge.

Ellis, G. (2001) Social exclusion, equality and the Good Friday Peace Agreement: the implications for land use planning, *Policy and Politics,* Vol. 29, No. 4, pp. 393-411.

Ellis, G. and McWhirter, C. (2002) *Good practice guide to promote racial equality in planning for Travellers - consultation document,* Belfast: Equality Commission for Northern Ireland.

Equality Commission for Northern Ireland (1999) *Guide to statutory duties,* Belfast: Equality Commission for Northern Ireland.

Eyben, K., Morrow, D. and Wilson, D. (1997) *A worthwhile venture? Practically investing in equity, diversity and interdependence in Northern Ireland,* Coleraine: University of Ulster.

Fenster, T. (1996) Ethnicity and citizen identity in planning and development for minority groups, *Political Geography,* Vol. 15, pp. 405-418.

Flora, C. (1997) Building social capital: the importance of entrepreneurial social infrastructure, *Rural Development News,* Vol. 21, No. 2, pp. 1-3.

Foley, M. and Edwards, B. (1999) Is it time to disinvest in social capital? *Journal of Public Policy,* Vol. 19, No. 2, pp. 141-173.

Friedmann, J. (1973) *Re-tracking America: a theory of transactive planning,* New York: Anchor Press.

Funder, A. (2003) *Stasiland,* London: Granta Books.

Gailey, A. (1984) *Rural houses of the North of Ireland,* Edinburgh: John Donald Publishers Ltd.

Gallagher, A. (1995) The approach of government: community relations and equity, in Dunn, S. (ed.) *Facets of the conflict in Northern Ireland,* pp. 27-43, New York: St Martin's Press.

Gebler, C. (1991) *The glass curtain: inside an Ulster community,* London: Abacus.

Graham, B. (1997) The imagining of place, in Graham, B. (ed.) *In search of Ireland,* pp. 192-212, London: Routledge.

Haase, T., McKeown, K. and Rourke, S. (1996) *Local development strategies for disadvantaged areas,* Dublin: Area Development Management Ltd.

Habermas, J. (1984) *The theory of communicative action- Vol. 1: Reason and the rationalisation of society,* London: Polity Press.

Habermas, J. (1987) *The philosophical discourse of modernity,* Cambridge: Polity Press.

Hart, M. and Murray, M. (2000) *Local development in Northern Ireland: the way forward,* Belfast: Northern Ireland Economic Council.

Harvey, B. (1994) *Combatting exclusion: lessons from the Third EU Poverty Programme in Ireland,* Dublin: Combat Poverty Agency.

Healey, P. (1996) Consensus-building across difficult divisions: new approaches to collaborative strategy making, *Planning Practice and Research,* Vol. 11, No. 2, pp. 207-216.

Healey, P. (1997) *Collaborative planning: shaping places in fragmented societies,* Basingstoke: Macmillan.

Healey, P. (1999) Institutional analysis, communicative planning and shaping places, *Journal of Planning Education and Research,* Vol. 19, No. 2, pp. 111-122.

Healey, P. (2003) Collaborative planning in perspective, *Planning Theory,* Vol. 2, No. 2, pp. 101-123.

Heikkila, E. (2001) Identity and inequality: race and space in planning, *Planning Theory and Practice,* Vol. 2, No. 3, pp. 261-275.

Higgins, G. and Brewer, J. (2003) The roots of sectarianism in Northern Ireland, in Hargie, O. and Dickson, D. (eds.) *Researching the Troubles: social science perspectives on the Northern Ireland conflict,* pp. 107-122, Edinburgh: Mainstream Publishing.

Hillier, J. (2003) Agonising over consensus: why Habermasian ideals cannot be real, *Planning Theory,* Vol. 2, No. 2, pp. 37-59.

Hoch, C. (1993) Racism and planning, *Journal of the American Planning Association,* Vol. 59, No. 3, pp. 451-460.

Home Office (2001) *Building cohesive communities: a report of the Ministerial Group on Public Order and Community Cohesion,* London: Home Office.

Howlett, M. (2000) Managing the hollow state: procedural policy instruments and modern governance, *Canadian Public Administration,* Vol. 43, No. 4, pp. 412-431.

Hughes, J., Knox, C., Murray, M. and Greer, J. (1998) *Partnership governance in Northern Ireland: the path to peace,* Dublin: Oak Tree Press.

Innes, J. (1998) Information in communicative planning, *Journal of the American Planning Association,* Vol. 64, No. 1, pp. 52-75.

Knox, C. and Quirk, P. (2000) *Peace building in Northern Ireland, Israel and South Africa: transition, transformation and reconciliation,* London: Macmillan Press Ltd.

Knox, C., Hughes, J., Birrell, D. and McCready, S. (1994) *Community relations and local government,* Coleraine: Centre for the Study of Conflict, University of Ulster.

Krishnarayan, V. and Thomas, H. (1993) *Ethnic minorities and the planning system,* London: Royal Town Planning Institute.

Lappe, F.M. and DuBois, P.M. (1997) Building social capital without looking backward, *National Civic Review,* Vol. 86, No. 2, pp. 119-128.

Liechty, J. (1993) *Roots of sectarianism in Ireland,* Paper commissioned for Working Party on Sectarianism set up by The Department of Social Issues of the Irish Inter-Church Meeting (Belfast).

Liechty, J. and Clegg, C. (2001) *Moving beyond sectarianism: religion, conflict and reconciliation in Northern Ireland,* Dublin: The Columba Press.

Local Government Information Unit (LGIU) (1995) *Race and regeneration: a consultation document,* London: LGIU.

Logue, K. (1992) *Anti-sectarianism and the voluntary and community sector,* Belfast: Community Relations Council.

Lowndes, V. and Skelcher, C. (1988) The dynamics of multi-organizational partnerships: an analysis of changing modes of governance, *Public Administration*, Vol. 76, Summer, pp. 313-333.

Lovan, W.R., Murray, M. and Shaffer, R. (eds.) (2004) *Participatory governance: planning, conflict mediation and public decision-making in civil society*, Aldershot: Ashgate.

Lovett, T., Gunn, D. and Robson, T. (1994) Education, conflict and community development in Northern Ireland, *Community Development Journal*, Vol. 29, No. 2, pp. 11-18.

MacGreil, M. (1996) *Prejudice in Ireland revisited*, Maynooth: National University of Ireland at Maynooth.

Manning-Thomas, J. (1994) Planning history and the black urban experience: linkages and contemporary implications, *Journal of Planning Education and Research*, Vol. 14, No. 1, pp. 1-11.

Mawson, J., Beazeley, M., Burfitt, A., Collinge, C., Hall, S., Loftman, P., Nevin, B., Srbljanin, A. and Tilson, B. (1995) *The Single Regeneration Budget: The Stocktake*, Birmingham: Centre for Urban and Regional Studies, University of Birmingham.

McMaster, J. (1993) *Young people as the guardians of sectarian tradition*, Belfast: Youth Link.

Miller, D. (ed.) (1998) *Rethinking Northern Ireland: culture, ideology and colonialism*, London: Longman.

Morrissey, M., McGinn, P. and McDonnell, B. (2002) *Report on research into evaluating community-based and voluntary activity in Northern Ireland*, Belfast: The Voluntary and Community Unit - Department for Social Development.

Morrow, D., Eyben, K., and Wilson, D. (2003) From the margin to the middle: taking equity, diversity and interdependence seriously, in Hargie, O. and Dickson, D. (eds.) *Researching the Troubles: social science perspectives on the Northern Ireland conflict*, pp. 163-181, Edinburgh: Mainstream Publishing.

Murray, M. and Greer, J. (1999) The changing governance of rural development: state-community interaction in Northern Ireland, *Policy Studies*, Vol. 20, No. 1, pp. 37-50.

Murtagh, B. (1996) *Community and conflict in rural Ulster*, Belfast: Northern Ireland Community Relations Council.

Murtagh, B. (1998) Community, conflict and rural planning in Northern Ireland, *Journal of Rural Studies*, Vol. 14, No. 2, pp. 221-231.

Murtagh, B. (2002) *The politics of territory*, Basingstoke: Palgrave.

New TSN Unit (1999) *Vision into practice: the first New TSN annual report*, Belfast: Corporate Document Service.

Noble *et al* (2001) *Measures of deprivation in Northern Ireland*, Belfast: Northern Ireland Statistics and Research Agency.

Northern Ireland Assembly, Public Accounts Committee (2000) *Report on the Rural Development Programme*, Belfast: The Stationery Office.

Northern Ireland Audit Office (2000) *The Rural Development Programme,* Belfast: The Stationery Office.

Northern Ireland Council for Voluntary Action (2002) *The state of the sector 111,* Belfast: Northern Ireland Council for Voluntary Action.

Northern Ireland Economic Council (2000) *Local development: a turning point,* Belfast: Northern Ireland Economic Council.

Northern Ireland Housing Executive (1990) *Rural housing policy review - leading the way,* Belfast: Northern Ireland Housing Executive.

Northern Ireland Housing Executive (1991) *Rural housing policy - the way ahead: a policy statement,* Belfast: Northern Ireland Housing Executive.

Oc, T., Tiesdell, S. and Moynihan, D. (1997) *Urban regeneration and ethnic minorities,* Bristol: Polity Press.

OECD (2001) *Engaging citizens in policy-making: information, consultation and public participation,* Paris: Organisation for Economic Cooperation and Development.

Office of the First and Deputy First Minister (2003) *A practical guide to policy making in Northern Ireland,* Belfast: Office of the First and Deputy First Minister.

Osborne, D. and Gaebler, T. (1992) *Reinventing government: how the entrepreneurial spirit is transforming the public sector,* New York: Plume.

Osborne, R. (1996) Policy dilemmas in Belfast, *Journal of Social Policy,* Vol. 25, No. 2, pp. 181-199.

Parker, S. (2001) Community, social identity and the structuration of power in the contemporary European city, part one: towards a theory of urban structuration, *City,* Vol. 5, No. 2, pp. 189-202.

Playboard (1994) *Gender matters: a guide to gender issues and children's play,* Belfast: Playboard.

Porter, N. (1998) *Rethinking Unionism: an alternative vision for Northern Ireland,* Belfast: Blackstaff.

PriceWaterhouseCoopers (1999) *Rural Development Programme mid term evaluation synthesis report,* Belfast: Department of Agriculture and Rural Development.

Putnam, R. (1993) *Making democracy work: civic traditions in modern Italy,* Princeton: Princeton University Press.

Putnam, R. (2000) *Bowling alone: the collapse and revival of American community,* New York: Simon and Schuster.

Qadeer, M. (1997) Pluralistic planning for multicultural cities: the Canadian practice, *Journal of the American Planning Association,* Vol. 63, No. 4, pp. 481-494.

Quirk, P. and McLaughlin, E. (1996) Targeting Social Need; in McLaughlin, E. and Quirk, P. (eds.) *Policy aspects of employment equality in Northern Ireland,* pp.153-185, Belfast: The Standing Advisory Commission on Human Rights.

Ramutsindela, M. (2001) Down the post-colonial road: reconstructing the post-apartheid state in *South Africa, Political Geography*, Vol. 21, pp. 57-84.

Remer, G. (2000) Two models of deliberation: oratory and conversation in ratifying the Constitution, *The Journal of Political Philosophy*, Vol. 8, No. 1, pp. 68-90.

Rhodes, R. (1997) Understanding governance: *policy networks, governance, reflexivity and accountability*, Buckingham: Open University Press.

Riley, F. (1994) Monitoring and race equality planning, in H. Thomas and V. Krisharayan (eds.) *Race and planning: policies and procedures*, Aldershot: Avebury.

Robson, B., Bradford, M. and Deas, I. (1994) *Relative deprivation in Northern Ireland*, Occasional Paper No. 28, Belfast: Policy Planning and Research Unit, Department of Finance and Personnel.

Roulier, M. (2000) Reconnecting communities and their schools through authentic dialogue, *National Civic Review*, Vol. 89, No. 1, pp. 53-65.

RTPI/CRE (Royal Town Planning Institute/Campaign for Racial Equality) (1983) *Planning for a multi-racial Britain: report of the Royal Town Planning Institute/Commission for Racial Equality Working Party*, London: CRE.

Rural Community Network (2003) *Workbook on equity, diversity and interdependence (EDI) in rural society - addressing our history of hear no evil, see no evil, speak no evil*, Cookstown: Rural Community Network.

Rural Development Council (1997) *A sense of belonging in Cookstown and Western Shores Network, Fermanagh, East Down and Oakleaf*, Cookstown: Rural Development Council.

Rural Development Council (2003) *A picture of rural change 2003*, Cookstown: Rural Development Council.

Sandercock, L. (1998) *Towards cosmopolis: planning for multi-cultural cities*, Chichester: John Wiley.

Sandercock, L. (2000) *When strangers become neighbours, Planning Theory and Practice*, Vol. 1, No. 1, pp.13-30.

Seekings, J. (2000) Introduction: urban studies in South Africa after apartheid, *International Journal of Urban and Regional Research*, Vol. 24, No. 4, pp. 832-840.

Seitles, M. (1996) The perpetualisation of residential racial segregation in America: historical discrimination, modern forms of exclusion and inclusionary remedies, *Journal of Land Use and Environmental Law*, Vol. 14, No. 1, pp. 1-30.

Shirlow, P. (2001) Devolution in Northern Ireland/ Ulster/ the North/ Six Counties: delete as appropriate, *Regional Studies*, Vol. 35, No. 8, pp. 743-752.

Shortall, S. and Kelly, R. (2000) *Gender proofing CAP reform*, Cookstown: Rural Community Network.

Shucksmith, M. (2000) *Exclusive countryside? Social inclusion and regeneration in rural areas*, York: Joseph Rowntree Trust.

Shucksmith, M. (2001) *History meets biography: processes of change and social exclusion in rural areas*, Paper presented at the seminar 'Exclusion Zones: Inadequate Resources and Civic Rights in Rural Areas', Belfast: Queen's University.

Singh, B.R. (2001) Dialogue across cultural and ethnic differences, *Educational Studies*, Vol. 27, No. 3, pp. 341-355.

Smith, S. (1989) *The politics of race and residence*, Cambridge: Polity Press.

Soja, E. (1999) In different spaces: the cultural turn in urban and regional political economy, *European Planning Studies*, Vol. 7, No. 1, pp. 65-75.

Soja, E. (2000) *Postmetropolis: critical studies of cities and regions*, London: Blackwell.

Symons, L. (1963) *Land use in Northern Ireland*, London: University of London Press.

Taylor, M. (1998) Combating the social exclusion of housing estates, *Housing Studies*, Vol. 13, No. 6, pp. 819-832.

Tewdwr-Jones, M. and Allmendinger, P. (1998) Deconstructing communicative rationality: a critique of Habermasian collaborative planning, *Environment and Planning A*, Vol. 30, pp. 1975-1989.

The General Synod of the Church of Ireland (2003) *The hard gospel: dealing positively with difference in the Church of Ireland – a scoping study report to the Sectarianism Education Project*, Belfast: The Church of Ireland Press Office.

Theakston, K. (1998) New labour, new Whitehall, *Public Policy and Administration*, Vol. 13, No. 1, pp. 13-34.

Thomas, H. (1997) Ethnic minorities and the planning system: a study revisited, *Town Planning Review*, Vol. 68, No. 2, pp. 195-211.

Thomas, H. (2000) *Race and planning: the UK experience*, London: UCL Press.

Turok, I. (1994) Urban planning in the transition from apartheid, *Town Planning Review*, Vol. 65, No. 4, pp. 243-259.

Yiftachel, O. (2000) Social control, urban planning and ethno-class relations: Mizrahi Jews in Israel's 'Development Towns', *International Journal of Urban and Regional Research*, Vol. 24, No. 2, pp. 418-437.

Yiftachel, O. (2001) Can theory be liberated from professional constraints? On rationality and explanatory power in Flyvbjerg's 'Rationality and Power', *International Planning Studies*, Vol. 6, No. 3, pp. 251-255.

Young, I.M. (1996) Communication and the other: beyond deliberative democracy, in Benhabib, S. (ed.) *Democracy and difference: contesting the boundaries of the political*, pp.120-135, Princeton: Princeton University Press.

Youth Action Northern Ireland (1997) *A sense of belonging*, Belfast: Youth Action Northern Ireland.

Index

aging population, 47
agonism, 131-3
agriculture. *see* Department of Agriculture; farming
Ahmed, Y. and Booth, C., 12
Albrechts, L., 121
alienation, 5, 7, 132
Amin, A., 135
anti-sectarianism, 19, 25, 61
Antrim (county), 34, 35, 52
area-based partnership governance, 29, 40
area-based service organisations, perceptions of, 81-104
Area-Based Strategies, 40
Aristotle, 123
Armagh/Monaghan borderlands, 34
Armstrong, J., McClelland, D. and O'Brien, T., 33-4
Arnstein, S., 122
assimilation model, 9-10
Atkinson, R., 17
authentic dialogue, 3, 6, 7, 119, 123-9
    concerted approach, need for, 126
    differential outcomes, 125-6
    dimensions of, 125
    EDI, application to, 125, 126
    social capital, and, 127-9
    storytelling, 123-5

Banbridge and Newry and Mourne Area Plan 2015, 83
Belfast, 37
    peacelines, 36
    population, 29, 30, 32
Belfast City region, 32, 35
Belfast Travel to Work Area, 83
Belfast Urban Area, 33-4
Best Value, 99, 100, 102, 137
Birrell, D. and Murie, A., 37
Bollens, S., 9, 10, 67
bonding social capital, 128
Booher, D. and Innes, J., 6, 7
border areas, 34
    Protestant communities in, 52
Bradford, 9

bridge building, 60
bridging social capital, 128
Bringing Britain Together: A National Strategy for Neighbourhood Renewal (1998), 16
Brooke, Peter, 23, 38
Bryan, D., 19
BSE, 84, 89
Burbules, N.C., 125-6
Burnley, 9

Caldwell, J. and Greer, J., 34
CAP reforms, gender proofing of, 36, 105
Catholic community, 9, 10, 20, 35. *see also* GAA
    community development, 52, 53
    rural development, and, 106
Central Area Local Plan, 13
Central Community Relations Unit (CCRU), 21, 103, 137
challenges and perspectives in EDI, 43-104, 112-15
    area-based service organisations, 81-104
    Dignity at Work (Newry and Mourne DC), 100, 101
    District Councils, 96-104
    District Partnerships, 91-5
    NI-wide service organisations, 43-80
    Rural Community Network (RCN), 106-10
    Sub-Regional Rural Support Networks, 81-7
Chicago Housing Authority, 11
children
    abuse of, 83
    Playboard training initiatives, 70-1
    rural Northern Ireland, 51-2
Christian religion
    sectarianism, and, 19
Christopher, A.J., 9
Church of Ireland
    The Hard Gospel (2003), 15
    Sectarianism Education Projects Committee, 15

Cincinnati Metropolitan Housing Authority, 11
citizenship
  alienation, 5
  dialogue with, 119-26. *see also*
authentic dialogue; consultation
  participatory, 38, 121, 122
  rights, 8, 9
  rural citizenry, nurturing of, xi
  social rights of, 16
City Challenge areas
  ethnic minorities, experiences of, 12
civic identity, 9
civic leadership, 59, 75, 101
civil society, 5, 38, 81, 121
civil unrest, 20
class realignment, 10
Cloke, P., 29
closure, sense of, 84-6
Coleraine Borough Council, 96, 99, 103-4
collaborative planning, xi, 3, 4-8, 10, 129, 131-5. *see also*
  authentic dialogue
  agonism, 131-3
  EDI, and, 4-8, 131-5
  limitations, 134
  value of EDI, 133
colour blind policy, 10, 67
comfort zones, 84
Commission for Racial Equality (CRE), 13, 61, 62, 137
Commission for Racial Equality (UK), 20
Common Agricultural Policy reforms
  gender proofing, 36, 105
communicative planning, 4, 6, 7
communicative theory, 4
community attitudes, 36-7
community development, 21, 29, 37-8.
  see also rural development
  Orange Order, 54-5, 85, 132-3
  Protestant community, 43, 52-5, 83, 84, 132-3
  Protestant perceptions, 43-4
  religious differences and, 52
  Rural Community Network (RCN), 105
  rural Northern Ireland, 37-9, 43, 52, 64
community dialogue, 73-7
Community Foundation for Northern Ireland, 143
community groups, 5, 37-8

Community Regeneration and Improvement Special Programme (CRISP), 40
community relations, 22, 24, 36, 58, 60, 63, 65, 102, 103, 115. *see also* cross-community initiatives
  District Councils, work of, 102-3
  EDI, and, 58-61, 64-5, 72, 74-5, 83-4, 115
  Education for Mutual Understanding (EMU), 21
  LEADER 2 Local Action Groups, 89-90
  LEADER 2 perspectives, 89-90
  legislation, 21
  local authorities programme, 21
  perceptions of, 58-9, 64-5
  play sector, in, 51-2
  public policy, 20-2
  Sub-regional Rural Support Networks, perspectives of, 83-7
  tokenism, 58-9, 64-5, 93-4
  traditional interpretations of, 85
  training, 52, 71-3
Community Relations Commission, 20-1
Community Relations Council (CRC), 10, 21, 25, 67-8, 71, 137
  Strategic Plan 1998-2001, 24
Community Relations Forum (Newry and Mourne), 74
Community Relations Officers, 58, 103
Community Relations Programme, 21
Community Relations Training and Learning Consortium (CRTLC), 71-3, 138
community renewal, 50
community seclusion, 84-6
community transport, 134
confidence building, 95
confidentiality, 113
conflict ('troubles'), 9, 14, 20, 51-2, 57-8, 84, 134. *see also* peace process
  British state management of, 9-10
  ceasefires (1994), 22
  ex-prisoners, 56
  victims and survivors, 56-8
Connolly, P., 19
consensus, 131
consultation, 15, 119, 120-2
Consultation Institute, 120
conversational cooperation, 123
Cookstown and Western Shores Network (CWSN), 81, 82, 84, 85

Cornwall, A., 122
Counteract, 25, 96, 106, 114, 138
   Gaining from Difference initiative, 106
Cousins, C., 16
Craigavon (Co. Armagh), 30, 45, 65
Craigavon Borough Council, 96, 101
CRC. *see* Community Relations Council
  Credit Union movement, 53
CRISP projects, 40
Cross-Border Action Group, 49
cross-border bodies, 144
cross-border networking, 105
cross-community initiatives, 20, 21, 64, 115
   tokenism, 58-9, 64-5, 93-4
cultural activities
   young people, involvement of, 50-1
cultural diversity, 22
Cultural Heritage, 21
cultural institutions. *see* GAA; Orange
  Order
cultural landscape, 32
cultural segregation, 132

DANI. *see* Department of Agriculture
dark side of difference, xi, 3, 132
De Souza Briggs, X., 127
Department of Agriculture (DANI), 35,
  38, 40
Department of Agriculture and Rural
Development, 39, 45, 79, 105, 124-5, 138,
142
Department of the Environment, 29, 30, 32
  CRISP projects, 40
Departmental Action Plans, 24
Derry/Londonderry, 29, 30, 35, 37, 45
devolved government, 22
DIAD model (*Diversity, Interdependence
and Authentic Dialogue*), 6, 7
dialogue, 5, 8, 73, 119, 123, 131, 134. *see
also* authentic dialogue; consultation
  communicative planning, 4, 6, 7
  inequalities in, 125
  leadership, 75
  research overview, 3-26
  rural people, 73-7
difference. *see also* diversity; religious
  difference
  dark side of, xi, 3, 132
  recognition of, 25
  social class, 85-6
  tolerance of, 24

Dignity at Work (Newry and Mourne DC),
  100, 101
direct rule, 20, 22, 53, 133, 138
disability, people with, 48
Disability Council, 139
discrimination, 8, 9
  fair employment, 22
  Fair Employment Tribunal, 22
  institutional, 44-5, 68
  political or religious prejudice, 19. *see
also* sectarianism
  racial, 11, 45
  Traveller community, against, 15, 44-5
discursive planning, 3-14, 134. *see also*
  collaborative planning
District Councils, 21, 78, 81, 96, 114, 138-9
  community relations work, 102-3
  distribution map, 97
  EDI perspectives, 96-104
  equality agenda, 99-100
  LEADER 2 local action groups, and, 40
  local economic development
partnerships, 40
  political polarisation, 102
District Partnerships, 40, 81, 93, 139
  cross-community "fixing," 93-4
  distribution map, 92
  EDI perceptions, 91-5
  establishment of, 93
  interdependence, 95
  social inclusion, promotion of, 93-4
  training, importance of, 94
diversity, 7, 51. *see also* EDI; rural
  diversity
  dealing with, 36-7
  definition, 24
  discourses on, 8-10
  respect for, 24, 36, 85
*Diversity, Interdependence and Authentic
  Dialogue* (DIAD), 6, 7
divided society, xi, 3, 36-7, 37, 38, 68
  collaborative planning, contribution of,
  3, 4-8
domestic violence, 83
Dorsett, R., 11
Down (county), 34, 35, 52
Duffy, P., 32

economic disparities, 11, 83
EDI, xiv, 24-6, 67, 68
  application of, 26

authentic dialogue. *see* authentic
dialogue
   challenges and perspectives. *see*
challenges and perspectives
   change in value systems, facilitation of,
61
   civic leadership, as, 59
   commitment to, 59
   community dialogue, 73-7
   community relations, and. *see*
community relations
   concept of, 58
   definitions, 24-5
   derivation of principles, 24
   inter-generational equity, 26
   inter-organisational collaboration, 77-80
   leadership, 75
   learning experience, as, 90
   legislative and policy superstructure, 25
   long-term nature of, 61, 113
   need for, 65
   obstacles to development of, 73, 84-7
   operationalisation of, 79
   potential of, 68
   principles, 24-5
   public policy, relevance for, 65
   rhetoric, 90-1, 123
   situation analysis, 111-15
   training needs, 68-9
   understanding of, 58-61
EDI Framework, 25
Education for Mutual Understanding
   (EMU), 21
Education Reform (Northern Ireland) Order
   1989, 21
elderly, 47
Ellis, G., 22, 35
Ellis, G. and McWhirter, C., 15
EQUAL, 144
equal opportunities, 24
equality, 20, 22-3, 61-6, 72, 133. *see also*
   equality agenda
     EDI, and, 61-6, 102
     opportunity, of, 10, 23
     training, 23, 66-7
     training in, 66
     urban-rural differentials, 66
equality agenda, 22-4, 62. *see also* equality
   impact assessments; equality schemes

additional workload, 64
District Councils, experiences of,
   99-101
   Promoting Social Inclusion (PSI), 24
   rural proofing, 66
   Targeting Social Need (TSN), 23-4
Equality Commission for Northern Ireland,
   36, 61, 62, 63, 64, 99
   functions, 139
   Good Practice Guide to Promote Racial
   Equality in Planning for Travellers, 15
equality impact assessments (EQIAs), 23,
   62, 99, 102, 139
equality schemes, 23, 47, 61, 63, 64, 65,
   140
   rural proofing, and, 66
equity, 7, 11-14, 32, 37, 44. *see also* EDI
   definition, 24
   inter-generational, 26
   rural Northern Ireland, 32
   Traveller community, for, 59-60
ethnic diversity, respect for, 14
ethnic geography of NI, 35
ethnic minorities, 11-14. *see also* Traveller
   community
   economic marginalisation, 11
   *Ethnic Minorities and the Planning
   System* (RTPI study), 13
   impact of policy on, 12
   planning, and, 11-14
   residential segregation, 10, 11
   training and business support for, 12
   urban regeneration governance, and, 12
ethnicity, 8
ethnocentrism, 18
EU funding, 35, 40, 41, 56. *see also* EU
   Special Support rogramme for Peace and
   Reconciliation
EU LEADER programme, 40, 87, 89, 141.
   see also LEADER 2 Local Action Groups
EU Special Support Programme for Peace
   and Reconciliation, 40, 58, 89, 106, 139,
   140
   objectives of, 91-2
EU Structural Funds Monitoring
   Committee, 63
EU Structural Funds Northern Ireland
   Single Programme 1994-1999, 40
European Commission, 16, 38

European Programmes Office, 94
evaluation, 114
ex ante appraisal, 23
ex-prisoners, 56-7
Eyben, K., Morrow, D. and Wilson, D., 25

Fair Employment Commission, 22, 139
Fair Employment (Northern Ireland) Act
  1989, 22
Fair Employment Tribunal, 22
fair treatment, 22
farmers' unions, women in, 46
farming, 33, 35, 84, 96
  aging population, 47
  CAP reforms, gender proofing of, 36
  crises of confidence, 84
  LEADER 2 Local Action Groups, 89
  social exclusion, 45
  women in, 45-6
Fenster, T., 8
Fermanagh (county), 35
Flags and Emblems Act 1989, 22
Flora, C., 127
Foley, M. and Edwards, B., 127
Foot and Mouth Disease, xii, 84, 89
France, 16
Friedmann, J., 121
Funder, Anna, *Stasiland* (2003), 129
funding, 53, 54, 62. *see also* EU funding
Future Ways Programme (University of
Ulster), 25, 89, 96, 114, 140
  Gaining from Difference initiative, 106

GAA (Gaelic Athletic Association), 50, 51,
55-6, 60, 80, 85, 94, 124, 132, 140
  role and contribution of, 55-6
  social inclusion, and, 56
Gaelic Athletic Association. *see* GAA
Gailey, A., 32-3
Gaining from Difference (Counteract/Future
  Ways), 106
Gallagher, A., 21
Games not Names (Playboard), 70
gay people, 47
Gender Matters (Playboard), 71
gender proofing, of CAP reforms, 36
General Election 2001, 35
General Synod of the Church of Ireland, 15
Germany, 129
global economy, 8

Good Friday Agreement 1998, 22, 23, 133,
  141, 144
*Good Practice Guide to Promote Racial
  Equality in Planning for Travellers*
  (Equality Commission), 15
Government Departments. *see also*
  Department of Agriculture
    Equality Schemes, 23
    Promoting Social Inclusion (PSI), 24
    Targeting Social Need (TSN), 23
Graham, B., 35
Grand Orange Lodge of Ireland, 143
Great Britain. *see* United Kingdom
group interactions, 18-19

Haase, T., McKeown, K. and Rourke, S., 34
Habermas, J., 4, 123
*The Hard Gospel: Dealing Positively
  with Difference in the Church of Ireland*
  (2003), 15
Hart, M. and Murray, M., 41, 81
Harvey, B., 17
Healey, P., 3, 4, 5, 8, 131, 132
  Institutional Audit, 5
Hekkila, E., 7
Higgins, G. and Brewer, J., 19
Hillier, J., 131
Hoch, C., 13
Home Office, 9
homelessness, 56
homophobia, 18
homosexuality, 47
housing
  affordability, 83
  homelessness, 56
  market pressures, 83
  redevelopment, 37
  rural, 32-3, 83
  Traveller community, discrimination
  against, 44-5
Howlett, M., 121
Hughes, J., Knox, C., Murray, M. and
  Greer, J., 20

identity, 9, 24, 55, 112, 132. *see also* single
  identity
inclusion. *see* social inclusion
Innes, J., 7
institution building, 4
institutional approach, 4

Institutional Audit, 5
institutional discrimination
    Traveller community, towards, 44-5, 68
institutional theory, 4
institutionalised equality agenda, 22
institutionalised segregation
    education, in, 21
integrated education, 10, 21
integration, 83-4, 89
inter-community differences. *see* religious
    difference
inter-denominational projects, 55
Inter-Departmental Committee on Rural
Development (IDCRD), 38, 39
inter-generational equity, 26
inter-organisational collaboration, 77-80
interactions, 18-19
interdependence, 7, 10-11, 15, 95, 134. *see
    also* EDI
    definition, 24-5
    local engagement, 26
    Traveller community, for, 59-60
interface management, 50
International Fund for Ireland (IFI), 40, 141
INTERREG 3A, 144
intra-community differences, 55, 86
Irish-Americans, 22
Israel, 9

Knox, C. and Quirk, P., 21
Knox, C., Hughes, J., Birrell, D. and
    McCready, S., 21
Krishnarayan, V. and Thomas, H., 13

Labour Government (Great Britain), 16
labour market, 17, 22
land
    meaning attached to, 96
    transfers across religious divide, 84
Lappe, F.M. and Du Bois, P.M., 127
LEADER+, 40, 141, 144
LEADER 2 Local Action Groups (LAG),
    40, 78, 81, 87, 145
    contribution of, 89-90
    distribution map, 88
    EDI challenges and perspectives, 87-91
    group relationships, 87, 89
    reconciliation, contribution to, 89-90
leadership, 59, 75
    training in, 101

Liechty, J., 19
Liechty, J. and Clegg, C., 19, 24
life worlds, 4
Limavady Borough Council, 96
linking social capital, 128-9
local authorities. *see also* District Councils
    Community Relations Programme, 21
    partnership governance, and, 40
Local Government Information Unit
(LGIU)
    Race and Regeneration (1995), 11-12
local partnerships, 39-40. *see also* District
    Partnerships
Local Strategy Partnerships (LSP), 40, 93,
    94, 141
Logue, K., 19
Londonderry (county), 35. *see also*
    Derry/Londonderry
Lovan, W.R., Murray, M. and Shaffer, R.,
    121
Lovett, T., Gunn, D. and Robston, T., 37
Lowndes, V. and Skelcher, C., 121
Loyal Orange Institution. *see* Orange Order

MacGreil, M., 18
McMaster, J., 19
Making Democracy Work (Putnam), 127
male suicides, 47, 83
Manning-Thomas, J., 13
marching bands, 50
marginalisation, 29
material citizenship, 10
Mawson, J. et al (1995), 12
Mediation Network, 74, 142
Miller, D., 19
mobility programmes, 11
modernisation, 10
monitoring, 114
Morrissey, M., McGinn, P. and McDonnell,
    B., 128
Morrow, D., Eyben, K. and Wilson, D.,
    21-2, 25-6
multi-nationalism, 8
multiculturalism, 8, 13
Murray, M. and Greer, J., 105
Murtagh, B., 10, 36, 132

narrative, 123-5
negotiating skills, 134
neighbourhood partnerships, 4

network power, 6
New TSN (NTSN), 23-4, 99, 142
  rural areas, 35, 65, 66
Newry and Mourne District Council, 96,
  99-101
  assessment of EDI, 103-4
  Community Relations Forum, 74
  Dignity at Work, 100, 101
  REDI group, 99
  voluntary code of conduct, 59
Noble et al, 2001, 35
non-sectarianism, 19, 61
Northern Ireland. *see also* rural Northern
  Ireland
  ethnic geography, 35
  geography, 31 (*fig.*)
  service organisations, perceptions of,
  43-80
  'troubles'. *see* conflict
Northern Ireland Act 1998, 22-3, 36, 139,
  140, 143
Northern Ireland Agricultural Producers
  Association (NIAPA), 43, 142
Northern Ireland Assembly, 22
  Public Accounts Committee, 41
  regional development strategy, 41
Northern Ireland Audit Office, 41
Northern Ireland Council for Travelling
  People, 144
Northern Ireland Council for Voluntary
  Action, 37
Northern Ireland Economic Council, 41
Northern Ireland Housing Executive
  (NIHE), 142
  rural housing policy review, 33
  Traveller community, and, 44
  The Way Ahead (1991), 33
Northern Ireland Rural Community
  Network (RCN). *see* Rural Community
  Network
Northern Ireland Rural Development
  Programme (RDP), 142
Northern Ireland Voluntary Trust (NIVT),
  57, 143
  REAL programme, 68-9

Oakleaf Network, 81, 82
obstacles, to development of EDI, 73,
  84-7, 134
Oc, T., Tiesdell, S. and Moynihan, D., 12

OECD, 119
Office of the First and Deputy First
  Minister, 120
Oldham, 9
Open College Network, 143
open countryside, 30, 32
Orange Order, 52, 53-4, 60, 80, 85, 86,
  94, 124, 132, 143
  community development, 54-5, 85,
  132-3
Osborne, D. and Gaebler, T., 121
Osborne, R., 22
overcrowding, 34

PAFT. *see Policy Appraisal and Fair
  Treatment*
paramilitarism, 19, 76, 128, 134
paramilitary ceasefires (1994), 22
Parker, S., 3, 7, 131
participative democracy, 38, 81
participatory citizenship, 38, 121-2, 122
partnership governance, 29, 39-41, 81.
  see also District Partnerships; Local
  Strategy Partnerships
  common characteristics, 40-1
  examples of, 40
  LEADER 2 Local Action Groups, 87-91
  local authorities, involvement of, 40
  rural development, and, 39-41
  sustainability, 41
PEACE 2, 144
peace building, 85, 89, 105
peace process, 23
  community relations, and, 83-4
  economic growth, 83
peace walls, 52
peacelines, 36
peripherality, 34-5
physical planning. *see* planning *A Picture
  of Rural Change* (RDC, 2003), 15
place dis-embeddedness, 3
place quality, 132, 134
planning, 3, 4-8, 30. *see also* collaborative
  planning
  colour blind policies, 10, 67
  communicative, 4, 6, 7
  discursive, 3-14
  diversity, and, 8-10
  divided society, and, 67-8
  ethnic impacts, monitoring of, 12

ethnic minorities, and, 11-14, 15
ethno-spatial context, 13
participatory approach, 121-2
race, and, 7, 11-14
racial advisors, 12
residential segregation, and, 10-11
stakeholder consultation, 120
training, 134
Travellers community, and, 15, 65
Playboard, 51, 143
    Games not Names, 70
    Gender Matters (1994), 71
    Play Without Frontiers initiative, 70
    training programmes, 70-1
pluralism, 8, 9, 19
policy. *see* public policy
Policy Appraisal and Fair Treatment
    (PAFT), 22, 23, 63, 65, 143
political conflict. *see* conflict
political divisions, 112
political identity
    single identity, and, 112
political prejudice, 18, 19. *see also*
    sectarianism
political support, 114
political symbols, public display of, 22
political transformation, 35
population, 30, 32
Porter, N., 5
poverty, 16-17, 134. *see also* social
    exclusion
prejudice, 18, 26. *see also* social prejudice
    awareness and reduction training, 66-73,
    91
    Travelling Community, to, 44
Price Waterhouse Coopers, 41
principles of EDI, 24-5
prisoners, 56-7
professional competency, 13
Promoting Social Inclusion (PSI), 24
Protestant community, 10, 20. *see also*
    Orange Order
    alienation, 132
    community development, 43, 52-5,
    83, 84
    community infrastructure, 52, 83, 132-3
    disintegration of rural communities, 52
    siege mentality, 54
    social inclusion projects, 55
    support, need for, 54-5

under-funding, 53, 54
    victims' groups, 57
provider-consumer interactions, 25
Public Accounts Committee (NIA), 41
public expenditure, 23
    Targeting Social Need (TSN), 23-4, 35
public funding dependency on, 40
public organisations
    authentic dialogue with, 123-6
    consultation, 120-2
public policy, xi, 20
    authentic dialogue, 119, 123-9
    community relations, 20-2
    EDI Framework, 25
    EDI principles, relevance of, 65
    equality agenda, 22-4
    fair treatment, 22
    features of good policy-making, 120
    Policy Appraisal and Fair Treatment
    (PAFT), 22, 23, 63, 65, 143
    policy proofing, 23
    stakeholder consultation, 119, 120-2,
    123
Public Voices Conference 1999
    (Craigavon), 101
Putnam, Robert, 4-5, 38, 127, 128

Qadeer, M., 8
Queen's University Belfast
    School of Environmental Planning, xii
Quirk, P. and McLaughlin, E., 23, 35

Race and Regeneration (LGIU), 11
Race Equality Progress Reports, 12
race relations, 20. *see also* ethnic
    minorities; racism
    planning, and, 7, 12-14
    race advisors, 12-13
    racism awareness training, 13
    sensitivity, 13
    training, 13
Race Relations Act, 13, 139
    Travellers, inclusion of, 44
race riots, 9
racial discrimination, 10
racial inequality, 7
racism, 11, 13, 45
Ramutsindela, M., 9
REAL programme, 68-9
reconciliation, 19, 25, 85, 89, 93

LEADER, contribution of, 89-90
   social inclusion and, tension between, 93
Regeneration of Mournes Area Ltd (ROMAL), 81, 82
regional development, 41
regional towns, 30
Relationships in Equality, Diversity and Interdependence (REDI), 99
religious difference, 86, 112
   Church of Ireland *Hard Gospel*, 15
   community development, and, 52, 53-4
   exclusion and, 52-3
   resistance to issues around, 134
   social exclusion, and, 52-4
religious prejudice, 18, 19. *see also* sectarianism
Remer, G., 123
Republic of Ireland, 34
research approach, xii-xiv
residential segregation, 11, 84
respect for diversity, 24, 36, 85
rhetoric, 90-1, 123
Rhodes, R., 121
Riley, F., 12
Robson, B., Bradford, M. and Deas, I., 35
Royal Town Planning Institute (RTPI), 13
   Ethnic Minorities and the Planning System, 13
Rural Action Project, 39
rural citizenry, nurturing of, xi
rural community, 30, 32
Rural Community Network (RCN), xii, xiii, xv, 36, 41, 57, 64, 91, 105, 114, 124, 128, 142, 144
   aims and objectives, 105
   EDI commitment, 78-9, 106
   EDI experience, 79-80, 106-10
   establishment, 39, 105
   membership, 105
   multi-level connections, 78
   perceptions of, 43-4, 106-8
   pivotal position of, 78
   sectarianism training, 108-9
   self-analysis, 108
rural development, 35, 37-9, 43-4, 60, 66, 67, 81, 91. *see also* Rural Community Network; Rural Development Council
   achievements, 39
   Area Based Strategy Groups, 40

DANI programme, 38, 39, 40, 41
   equality, 'rural proofing' and, 66
   farmers and, 45
   Inter-Departmental Committee (IDCRD), 38, 39
   LEADER 2 Local Action Groups (LAG), 40, 81, 87-91, 144
   partnership governance and, 39-41
   sectarianism, and, 89-90
Rural Development Council (RDC), 41, 79, 80, 142, 144
   establishment, 38
   A Picture of Rural Change (2003), 15
   A Sense of Belonging (1997), 84-5
   Strategic Plan 1992-1995, 38-9
Rural Development Division, 45
Rural Development Initiative, 38-9
rural disadvantage, 33-5, 66, 106
rural diversity, 33-4, 36-7
   area-based service organisations, perceptions of, 81-104
   NI-wide service organisations, perceptions of, 43-80
rural economy, 83, 84. *see also* farming
Rural Enterprise Division, 45
rural exclusion, 43-58, 111. *see also* social inclusion
   children, 51-2
   community perceptions, organisational reaction to, 60-1
   elderly, 47
   ex-prisoners, 56
   farmers, 45, 47
   housing, 44, 56
   intra-denominational tension, 55
   manifestations of, 44-8
   religious difference, 52-6, 84-5
   sexual orientation, 47
   social class differences, 85-6
   Traveller community, 44-5
   victims and survivors, 56-8
   women, 45-6
   young people, 47-52, 83
rural housing. *see* housing
rural imagery, 32-3
Rural Interface Initiative, 49
rural Northern Ireland, xi, 29-115.
   see also rural society definition, 29-30
   EDI challenges and perspectives, 37, 43-104

governance arena, 29-42
hard infrastructure, 53
open countryside, 30, 32
population, 30, 31, 32, 52
problem areas, 34
scale, 29-32
small towns, 30, 32
soft infrastructure, 53
villages, 30, 32
rural peripherality, 34-5
rural population, 30, 32
rural problem areas, 34
rural proofing
    equality agenda, 66
    meaning of, 144
rural scale, 29-32
rural society, 111
    aging population, 47
    categories of rurality, 33-4
    children in, 51-2
    closeness and closure, 84
    community attitudes, 36-7
    community development, 29, 37-9
    community dialogue, 73-7
    community relations, 83-5, 89-90
    community renewal, 50
    cultural landscape, 32-3, 84
    dealing with diversity, 36-7. *see also*
rural diversity
    development. *see* community
development; rural development
    discourse, limits on, 132
    divided society, 36-7
    inequalities, 32, 37, 111
    inter-organisational collaboration, 77-80
    isolation, 66
    restructuring of, 84
    segmentation, 111
    social class differences, 85-6
    social exclusion. *see* rural exclusion
    spatial segregation, 84-5
    women, 45-6
    young people, 47-52
Rural Support Networks (RSN), 105
rural-urban dialectic, 29

Sandercock, L., 3, 13
school closures, 52-3
School of Environmental Planning (QUB),
  xii

sectarianism, 19-20, 26, 36-7, 37, 48, 58,
70, 77, 89, 102, 105, 112, 134
    approaches for dealing with, 19-20
    Christian religion, role of, 19
    Church of Ireland *Hard Gospel* (2003),
15
    community dialogue, 73-7
    definition, 19
    manifestations of, 19
    moving beyond, 20
    rural areas, 48
    rural development, and, 89-90
    rural Northern Ireland, 36, 37
    social prejudice, 18-19
    training, 108-9
Seekings, 2000, 9
segregation, 19, 20
    education, in, 21
    residential, 10-11
    resistance to issues around, 134
    spatial, 10, 84-5, 132
    strategic reasons, 11
    tactical reasons, 11
Seitles, M., 11-12
A Sense of Belonging (RDC, 1997), 84-5
A Sense of Belonging (YANI, 1997), 47-8,
51
sexism, 18
sexual orientation, 18, 47
    exclusion on grounds of, 47
Sheffield
    Race Equality Progress Reports, 12
Shirlow, P., 10
Shortall, S. and Kelly, R., 36
Shucksmith, M., 16-17
siege mentality, 54
Singh, 2001, 125, 126
single identity
    political identity and, distinction
between, 112
    projects, 20, 21, 60, 85
Single Regeneration Budget (SRB), 11
    Bidding Guidance to Fourth Round, 12
Sinn Fein, 35
sites of resistance, 3, 7, 131
situation analysis, 111-15
skills development. *see* training
small farmers, 33
small towns, 30, 32
Smith, S., 11

social capital, 6, 51, 80, 112, 127, 132
  authentic dialogue and, 127-9
  components of, 128
  definition, 4
  development of, 4
  discursive planning and, 4-6
social class differences, 85-6
Social Democratic and Labour Party
  (SDLP), 35
social deprivation, 33-5
social disintegration, 16
social distance, 10
social exclusion, 15, 16-18, 26, 105, 131
  definition, 16-17
  forms of, 17
  hierarchy of attitudes, 63
  manifestations of, 44
  multiple system breakdown, 17
  new forms of, 83
  rural areas, 43-58. *see also* rural
exclusion
  Traveller community, 44, 124
social inclusion, 17-18, 20, 23, 25, 36, 40
  area-based service organisations,
  perceptions of, 81-104
  dealing with diversity, 36-7
  District Partnerships, contribution of,
93-4
  GAA, 56
  inclusionary zoning, 11
  NI-wide service organisations,
  perceptions of, 43-80
  Promoting Social Inclusion (PSI), 24
  reconciliation and, tension between, 93
  Targeting Social Need (TSN), 23-4, 35,
102
social need, 23-4
social prejudice, 18-19, 26, 37, 112
  definition, 18
  manifestations of, 18
  prejudice awareness and reduction
training, 66-73, 91
  sectarianism, and, 18, 36. *see also*
sectarianism
social rights of citizenship, 16
social segregation, 20
societal diversity, 15
societal transformation, 16, 26
Soja, E., 7, 131
South Africa

residential integration, 9
South Armagh, 40
  youth initiative, 48, 49
South Down
  housing market pressures, 83
South Down network (ROMAL), 81, 82
South Tyrone, 48
space, 7, 8, 132, 134
spatial planning, 30, 121. *see also* planning
spatial segregation, 10, 84-5, 132
Special European Programmes Body, 144
sporting activities
  Gaelic Athletic Association, 50-1, 55-6
  young people, involvement of, 50-1
stakeholder consultation, 120-2, 132
Standing Advisory Commission on Human
  Rights, 22, 144
Stasiland (Funder), 129
storytelling, 123-5
Sub Regional Rural Support Network, 66,
  78, 81, 114, 144
  community relations perspectives, 83-7
  distribution map, 82
  economic disparities, 83
  EDI perspectives, 81-7
  young people, needs of, 83
suicides, male, 47, 83
survivors, 56-7, 57
Symons, L., 33
system support, 17

Targeting Social Need (TSN), 23-4, 35,
  102, 133
Taylor, M., 4
tenure restructuring, 10
territorial conflicts, 135
territoriality, 10
terrorism, 40
Tewdwr-Jones, M. and Allmendinger,
  P., 3, 7
The Way Ahead (NIHE, 1991), 33
Theakston, 1998, 121
Third EU Poverty Programme in Ireland,
  17
Thomas, H., 8, 13
tokenism, 58-9, 64-5, 93-4
tourism networks, 89
training, 87, 90, 114, 134
  civic leadership, 101
  Community Relations Training and

Learning Consortium (CRTLC), 71-3
  District Partnership Boards, 94
  equality, 23, 66-7
  mutual respect, 68
  Playboard initiatives, 70-1
  prejudice awareness and reduction,
  66-73, 91
    racism awareness, 13
    REAL programme (NIVT), 68-9
    sectarianism training, 108-9
    Travellers' Movement anti-
  discrimination pack, 69
trauma, 57, 87
Traveller community, 15, 79, 144
  attitudes towards, 44-5
  cultural heritage, 45, 124
  discrimination against, 15, 44-5, 68, 69
  equality for, 15
  equity for, 59-60
  interdependence for, 60
  planning policy and, 15, 65
  prejudice towards, 18
  social exclusion, 44, 124
Travellers' Movement, 144
  anti-discrimination pack, 69
'troubles'. *see* conflict
trust, 4, 19, 134
TSN. *see* Targeting Social Need
Turok, I., 9
Tyrone-Armagh-Down-Antrim Network
  (TADA), 81, 82
Tyrone (county), 35

Ulster Farmers Union (UFU), 43, 144
understanding, 125, 126
unemployment, 16, 17, 23, 34
Unitary Development Plan, 13
United Kingdom
  *Bringing Britain Together: A National
  Strategy for Neighbourhood Renewal
  (1998), 16
  race riots, 9
  Social Exclusion Unit, 16
  Urban Programme, 11
United States
  inner-city redevelopment, 13
  mobility programmes, 11
  race and planning, 13-14
  residential segregation, 11-12
University of Ulster. *see* Future Ways
  Programme

URBAN 2, 144
Urban Development Corporations
  race issue, 12
Urban Programme, 11
urban regeneration governance structures
  ethnic minority representation, 12

vernacular dwellings, 33
victims and survivors, 56-8
villages, 30, 32
violence, 20, 36, 37
  legacy of, 83
voluntary sector, 5, 37, 39. *see also*
  community development; Rural
  Community Network (RCN)

The Way Ahead (NIHE, 1991), 33
way forward, 119-35
'West of the Bann' (slogan), 33, 106
Westminster, direct rule from, 20, 22,
  53, 138
women, 45-6
  violence against, 83
women's groups, 46
Women's Institute, 144
women's networks, 46

YANI. *see* Youth Action - Northern Ireland
Yiftachel, O., 7, 8-9
Young, I.M., 123-4
young people. *see also* children
  culture and sporting activities, 50-1
  difference, hierarchy of, 48
  males, suicides among, 47, 83
  social exclusion, 48-52, 83
Youth Action - Northern Ireland (YANI),
  47, 144
  Cross-Border Action Group, 49
  interface management, 50
  Rural Interface Initiative, 49
  Rural Model Projects, 49
  Rural Unit, 48
  A Sense of Belonging (1997), 47-8, 51
  South Armagh Youth Initiative, 48, 49
youth work
  rural areas, in, 48
youth workers
  Playboard training initiatives, 70-1

zero-sum politics, 5

Printed and bound by CPI Group (UK) Ltd, Croydon, CR0 4YY

22/10/2024

01777640-0005